Contents

Department and Agency Abbreviations

Department of Agriculture	USDA
Department of Commerce	DOC
Department of Defense	DOD
Department of Education	ED
Department of Energy	DOE
Department of Health and Human Services	HHS
Department of Homeland Security	DHS
Department of the Interior	DOI
Department of Transportation	DOT
Environmental Protection Agency	EPA
National Aeronautics and Space Administration	NASA
National Institute of Standards and Technology (part of DOC)	NIST
National Institutes of Health (part of HHS)	NIH
National Oceanic and Atmospheric Administration (part of DOC)	NOAA
National Science Foundation	NSF
National Science and Technology Council	NSTC
Nuclear Regulatory Commission	NRC
Office of Management and Budget	OMB
Office of Science and Technology Policy	OSTP
Smithsonian Institution	SI
United States Geological Survey (part of DOI)	USGS

1. Executive Summary

"One of the things that I've been focused on as President is how we create an all-hands-on-deck approach to science, technology, engineering, and math. We need to make this a priority to train an army of new teachers in these subject areas, and to make sure that all of us as a country are lifting up these subjects for the respect that they deserve."

President Barack Obama
2013 White House Science Fair
April 2013

Advances in science, technology, engineering, and mathematics (STEM) have long been central to our Nation's ability to manufacture better and smarter products, improve health care, develop cleaner and more efficient domestic energy sources, preserve the environment, safeguard national security, and grow the economy. For the United States to maintain its preeminent position in the world it will be essential that the Nation continues to lead in STEM, but evidence indicates that current educational pathways are not leading to a sufficiently large and well-trained STEM workforce to achieve this goal. Nor is the U.S. education system cultivating a culture of STEM necessary for a STEM-literate public. Thus it is essential that the United States enhance U.S. students' engagement in STEM disciplines and inspire and equip many more students to excel in STEM.

Investing in STEM education is critical to the Nation and its economic future for a number of reasons:

- The jobs of the future are STEM jobs: The demand for professionals in STEM fields[7] is projected to outpace the supply of trained workers and professionals. Additionally, STEM competencies are increasingly required for workers both within and outside specific STEM occupations. A recent report by the President's Council of Advisors on Science and Technology (PCAST) estimates there will be one million fewer STEM graduates over the next decade than U.S. industries will need.[8]

- Our K-12 system is "middle of the pack" in international comparisons: Among 33 Organization for Economic Cooperation and Development (OECD) countries that participated in a recent

Programme for International Student Assessment (PISA) study, which measures students' ability to apply what they have learned in reading, mathematics, and science and has been designed to assess whether students can use their knowledge in real-life situations[1], 12 countries had higher scores than did the United States in science and 17 had higher scores in mathematics.[9]

- Progress on STEM is critical to building a just and inclusive society: STEM participation and achievement statistics are especially disturbing for women and minorities, who are substantially underrepresented in STEM fields. While earning a STEM degree is one important milestone in pursuing a STEM career, just 2.2 percent of Hispanics and Latinos, 2.7 percent of African Americans, and 3.3 percent of Native Americans and Alaska Natives have earned a first university degree in the natural sciences or engineering by age 24.[10] While women constitute the majority of students on college campuses and roughly 46 percent of the workforce, they represent less than one in five bachelor's recipients in fields like computer science and engineering, and hold only 25 percent of STEM jobs.[11]

A Strategic Plan for Federal Investment in STEM Education

Many of the CoSTEM agencies have placed a high priority on STEM education and have developed education initiatives unique to their agency's mission, needs, and resources. To better leverage these assets and expertise, the Administration is releasing this STEM education strategic plan, the result of extensive cross-agency collaboration, to articulate a strategy for making progress on this national priority. The Administration, through the CoSTEM agencies, is committed to laying groundwork that will set the course for a coherent and impactful collective Federal STEM education investment for the next five years.

The Plan begins by providing an overview of the importance of STEM education to American scientific discovery and innovation, the need to better prepare students for today's jobs and those of the future, and the importance of a STEM-literate society (section 2) and also describes the current state of Federal STEM education efforts (section 3). The document then presents five priority STEM education investment areas where a coordinated Federal strategy can be developed, over five years, designed to lead to major improvements in key areas. This increased coordination is intended to lead to maximum impact and, as it is implemented, will lead to strategies for closer and more effective coordination among agencies with STEM investments (section 4).

Also included in this plan are initial implementation roadmaps in each of the priority STEM education investment areas, proposing potential short-, medium-, and long-term objectives and strategies that might help Federal agencies achieve the outlined goals (section 5). Additionally, throughout the document, the

---

[1] http://www.oecd.org/pisa/aboutpisa/

plan highlights (1) key outcomes for the Nation and ways Federal agencies can contribute, (2) areas where agencies will play a lead role, thereby increasing accountability, (3) methods to build and share evidence, and (4) approaches for decreasing fragmentation.

Choosing national goals that Federal agencies can contribute to

The STEM Strategic Plan sets out ambitious national goals to drive Federal investment in five[12] priority STEM education investment areas:

- Improve STEM Instruction: Prepare 100,000 excellent new K-12 STEM teachers by 2020, and support the existing STEM teacher workforce;
- Increase and Sustain Youth and Public Engagement in STEM: Support a 50 percent increase in the number of U.S. youth who have an authentic STEM experience each year prior to completing high school;
- Enhance STEM Experience of Undergraduate Students: Graduate one million additional students with degrees in STEM fields over the next 10 years;
- Better Serve Groups Historically Under-represented in STEM Fields: Increase the number of students from groups that have been underrepresented in STEM fields that graduate with STEM degrees in the next 10 years and improve women's participation in areas of STEM where they are significantly underrepresented; and,
- Design Graduate Education for Tomorrow's STEM Workforce: Provide graduate-trained STEM professionals with basic and applied research expertise, options to acquire specialized skills in areas of national importance, mission-critical workforce needs for the CoSTEM agencies, and ancillary skills needed for success in a broad range of careers.

STEM Education Coordination Approaches

Central to the success of this Strategic Plan is moving toward a new approach to coordinating Federal investments in STEM education (Section 4.2). By designating initial lead and collaborating agencies in some of the priority STEM education investment areas, the Strategic Plan encourages a more deliberative focus among new and existing efforts, the expansion of existing collaborations, and the creation of new synergies. The intent is to establish a coordinated, coherent portfolio of STEM education investments across the Federal Government so efforts and assets are deployed effectively and efficiently, for greatest potential impact. To do so, Federal agencies will focus on two main STEM education coordination approaches:

- Build new models for leveraging assets and expertise. Implement a strategy of lead and collaborating agencies to leverage capabilities across agencies to achieve the most significant impact of Federal STEM education investments.

- Build and use evidence-based approaches. Conduct STEM education research and evaluation to build evidence about promising practices and program effectiveness, to be used across agencies, and share with the public to improve the impact of the Federal STEM education investment.

Congressional leadership and commitment to STEM education stimulated the call for this plan and has been critical in its development. The main body of this report further describes the priority STEM education investment areas and STEM education coordination approaches, and provides initial implementation roadmaps to achieve these strategic objectives. The Administration, including the COSTEM agencies, looks forward to working with legislative leaders on its continued refinement and implementation.

2. Introduction

"We want to make sure that we are exciting young people around math and science and technology and computer science. We don't want our kids just to be consumers of the amazing things that science generates; we want them to be producers as well. And we want to make sure that those who historically have not participated in the sciences as robustly — girls, members of minority groups here in this country — that they are encouraged as well. We've got to make sure that we're training great calculus and biology teachers, and encouraging students to keep up with their physics and chemistry classes.... It means teaching proper research methods and encouraging young people to challenge accepted knowledge."

President Barack Obama
National Academy of Sciences
April 2013

Increasing opportunities for young Americans to gain strong STEM skills is essential if the United States is to continue its remarkable record of success in science and innovation. Numerous advances, from mapping the human genome to discovering water on Mars to developing the Internet, would not have been possible without a skilled and creative STEM workforce. New technologies and STEM knowledge lie at the core of our ability to manufacture better, smarter products, improve health care, preserve the environment, and safeguard national security. Individuals prepared with the skills and knowledge to invent, build, install, and operate those new technologies are essential. In addition, a basic understanding of STEM topics and concepts is necessary beyond the workplace in order for citizens to make informed decisions on issues that are increasingly at the center of local and national political debates, such as environmental regulation. STEM literacy is also critical when it comes to making sound personal consumer choices, from health-care decisions to purchases at the grocery store.

STEM knowledge and skills are in even greater demand as the United States confronts a fiercely competitive international marketplace where the advantage goes to companies that are the first to invent and produce innovative products. From 2000 to 2010, the growth in STEM jobs was three times greater than that of non-STEM jobs.[13] The Department of Commerce estimates that in the coming years STEM occupations will grow 1.7 times faster than non-STEM occupations.[14] Furthermore, Georgetown University's Center on Education and the Workforce projects that America will create 779,000 jobs between 2008 and 2018 that require a graduate degree in a STEM field but, based on current trends, only 550,000 native-born Americans will earn STEM graduate degrees during this period.[15]

Industry reports of inability to fill large numbers of STEM-related jobs and other workforce data raise concerns about U.S. capacity to meet the growing demand for STEM workers with appropriate training and skills.[16] The private and public sectors, including the Federal agencies, rely on the U.S. education system and outside-of-school learning opportunities to equip youth with fundamental knowledge and skills needed for choosing and training for their careers, and for making informed decisions as workers, parents, and members of the public.

Furthermore, recent reports indicate that:

- Students who report early expectations for a career in science are much more likely to complete a college degree in a STEM field than students without those expectations. This suggests that early exposure to science topics, at middle grades or below, may be important for a student's future career aspirations.[17]

- The achievement gap in mathematics and science remains a persistent issue. According to one recent report on international assessment of mathematics and science, the science scores of white U.S. eighth graders were surpassed only by the scores of three counties (Singapore, Chinese Taipei, and Korea), while Hispanic and black U.S. eighth graders had scores equivalent to those of students in countries ranked in the bottom third of the 45 countries that participated in the 8th grade science assessment.[18] Only one in five high school graduates who scores in the top quartile in mathematics goes on to become a STEM professional.[19]

- Fewer than 40 percent of students who enter college intending to major in a STEM field complete a STEM degree.[20]

- Only 19 percent of U.S. bachelor's degrees are awarded in STEM fields, while in China over 50 percent of first degrees are awarded in STEM fields.[21]

- Underrepresented minorities in STEM now account for almost 40 percent of K-12 students in the U.S.; however, they earn only 27 percent of all associate's degrees from community colleges, 17 percent of the bachelor's degrees in the natural sciences and engineering, and 6.6 percent of the doctorates in those fields.[22,23]

- Roughly 30 percent of chemistry and physics teachers in public high schools did not major in these fields and have not earned a certificate to teach those subjects.[24]

- Women make up nearly 50 percent of the U.S. workforce and a majority of college students, but hold less than 25 percent of STEM jobs[25] and earn less than one in five bachelor's degrees in high growth fields like computer science and engineering.[26]

Inadequacies in education pathways leading to STEM degrees and into the workforce amplify concerns that the United States is failing to keep pace with its international competitors in producing a workforce with the necessary skills and knowledge to advance STEM fields.

Given that accomplishing the complex missions of many of the CoSTEM agencies necessarily relies upon a strong and skilled STEM talent pool, those agencies have placed a high priority on trying to improve STEM education and workforce training and developing education and preparation initiatives unique to their respective missions, needs, and resources. *The Federal Science, Technology, Engineering, and Mathematics Education 5-year Strategic Plan* detailed here is intended to improve the efficiency, coordination, and impact of these federally supported STEM education investments. [27]

The Strategic Plan proposes approaches for Federal agencies to coordinate these activities and leverage these assets to make progress on the national priority of STEM education. To do so, the Strategic Plan identifies five priority STEM education investment areas and two coordination strategies for organizing Federal investments in STEM education. Through execution of the Strategic Plan, Federal agencies will collectively address critical STEM challenges with the intention of improving the impact of Federal STEM investments and making faster progress toward an ambitious vision. The Plan builds upon steps the Federal agencies have already taken to improve their STEM education plans and evaluation strategies and to coordinate their efforts[28][29][30] and takes the next step toward realizing the shared commitments of the Administration and the Congress to improve STEM education.

3. Federal Role in STEM Education

The 14 Federal CoSTEM agencies invest in and support programs and activities to improve STEM education. These agencies also maintain and fund such important assets as laboratories, research instruments, and facilities, and employ engaged and knowledgeable scientists, researchers, and engineers. Many of their programs are designed to develop a STEM-literate population and to ensure a highly qualified workforce in agency-related fields as well as in the STEM fields more generally.

To do so, they support all levels of learners and all learning environments, including preschool, K-12, two- and four-year colleges, universities, and informal-learning environments. Many programs and investments are also designed to provide learning resources for the general public, including publications, websites, television programs, museum exhibits, after-school programs, and video resources, among others. Federal support for STEM education currently enables:

- Preparation and professional development of STEM teachers and undergraduate faculty in their subject areas, in pedagogy, and in instructional practice through pre-service education and continuing professional development, as well as efforts to enhance teacher recruitment and retention;
- development of instructional materials, learning resources, and courses, including materials that can be integrated into curricula (such as videos, assignment and activity ideas, computer visualizations and simulations), and platforms for building and delivering interactive online courses and learning objects;
- training and re-training to match U.S. workforce skills to the demands of a rapidly-changing global economy and the STEM workforce to accomplish the missions of Federal agencies;
- direct support to students in disciplines related to agency missions through scholarships, fellowships, immersion research experiences in agency operations, training grants, internships, and other programs;
- research and development to understand and improve STEM education and learning programs at all levels, including research on: STEM learning and instructional strategies, learning in informal environments, ways to improve the preparation and professional development of teachers and faculty, STEM workforce development educational programs, and broadening participation education efforts;
- availability of facilities and staff to institutions engaged in STEM education;
- data collection initiatives and program evaluations; and
- public education and lifelong learning projects, including publications, websites, videos, simulations, television programs, museum exhibits, and public events.

Individual agency support of STEM education has evolved as a result of a combination of factors, including the mission and goals of the agency, statutory roles and responsibilities,[2] Congressional and Presidential directives to engage in particular aspects of STEM education, and the unique assets each agency has to contribute.

Most of the CoSTEM agencies support STEM education to meet their specific missions and workforce needs. Some also leverage their facilities, assets, technical workforce, and expertise to support fundamental STEM education research and development and to develop and implement programs that promote STEM literacy or proficiency in mission-related areas. The Federal agencies, including but not limited to the National Oceanic and Atmospheric Administration and the Environmental Protection Agency, that are responsible for maintaining natural resources and environments and for protecting life and property also support STEM education by making their spaces available to educational efforts. This is one of a number of strategies that Federal agencies use to improve stewardship of natural resources and informed decision making by the general public. Agency stewardship programs also typically include outreach, communication, and community engagement activities.

The majority of the Federal investment in STEM education – which totals approximately $3 billion annually (see Appendix A) – is funded largely through programs executed at the NSF, the Department of Education (ED), and, in the biomedical sciences, at the National Institutes of Health. NSF, within its mission to support basic science research and STEM improve education, invests in advancing knowledge through research and development of tested models about how to understand and improve STEM learning. This encompasses teaching and learning from the K-16 through post-graduate levels, as well as continuing education, retraining, and informal education in out-of-school settings. NSF also has a well-established graduate fellowship program that spans all areas of STEM except the biomedical sciences. NSF also has a history of partnering with a number of other CoSTEM agencies in areas of science and education.

ED supports programs to improve education in the United States and has a broad mission to promote student achievement and preparation for global competitiveness. Although only a small fraction of ED's funding supports specific STEM education programs, STEM initiatives have been a competitive priority in such significant programs as Race to the Top[31] and the Investing in Innovation Fund.[32] Given its mission, ED does not have substantial direct access to science and engineering research activities or to a STEM workforce through substantial in-house or external research and development (R&D) programs, but is developing approaches to partner effectively with the other CoSTEM agencies. ED brings unmatched reach to schools, teachers, and students across the Nation, and so this plan provides

---

[2] Several Federal agencies have statutory responsibility to provide STEM education. These include the Department of Commerce [DOC], Department of Energy [DOE], the Department of the Interior [DOI], Department of Defense [DOD], the National Aeronautics and Space Administration [NASA], and the U.S. Department of Agriculture [USDA].

approaches for leveraging these important connections. Furthermore, ED is building a staff with expertise in STEM teaching, and the National Center for Education Statistics cultivates data critical for STEM education research, including for efforts like the Institute of Education Sciences (IES) Mathematics and Science Education Research Grants Program.

The Smithsonian Institution (the Smithsonian), founded in 1846 and comprising19 museums and galleries, the National Zoological Park, and nine research facilities, promotes a number of engagement opportunities across a variety of disciplines. In 2012, the Smithsonian welcomed 30 million visitors to its public institutions and 103 million users to its website. In the context of Federal STEM activities, the Smithsonian has unique capabilities for serving as a clearinghouse that can reach a wide public audience.

Particular agencies, because of their mandates and goals, have specific expertise and assets to bring to the Federal STEM education investment in collaboration with the designated lead agencies. NASA has exciting assets associated with space exploration, and the USDA has networks that reach across the country through 4-H, extension services programs, and regional networks and laboratories. The NIH oversees traineeship and fellowship programs with a focus on graduate and post-doctoral research that prepare tomorrow's leading biomedical scientists. In addition, agencies such as NASA, HHS, DOD, DOC, DOE, and USDA support the teaching of mission-related science and engineering through connections to the scientific and technical assets that they oversee. Many of the CoSTEM agencies invest in or manage scientific, technological, engineering, and mathematics experts, research facilities and technology, data sets, and natural resources that can be leveraged as sites for STEM learning and workforce training. Some agencies (e.g., NSF, ED/Institute for Education Sciences) focus on building and gathering evidence about effective professional development practices or models, while others (e.g. ED) have mechanisms for scaling up effective practices or systemic changes in professional development.

This diversity of missions and approaches has over time led to an uncoordinated Federal investment in STEM education and STEM programs have proliferated to the point where in FY 2012, there were 226 programs across 13 different agencies described in Appendix A. This distributed approach to making critical investments in STEM education has made it difficult to ensure that Federal efforts are coherent, strategic, and leveraged for greatest impact. At the same time, the activities supported by the agencies have important functions and with coordination the combined efforts can unquestionably be greater.

Aside from STEM-specific education programs, the Federal Government also supports the development of STEM skills through investments that have goals that extend beyond STEM education (e.g., the many general education programs at ED), or those that focus on STEM research. At the postsecondary level, a portion of the Federal research and development (R&D) expenditure supports investigator-driven research awards at colleges and universities where research teams may include STEM-focused undergraduate students and graduate research assistants. While science R&D funding provides critical training opportunities for aspiring STEM graduates, and contributes to the development of a skilled workforce, such investments are outside the scope of STEM education considered in this Plan.

Although the Federal Government plays an important role in STEM education, it cannot achieve success by itself. To effectively leverage its investments, the Federal Government must coordinate its efforts strategically and collaborate with non-Federal partners to support institutional, state, and local efforts. Local and state education agencies, institutions of higher education, professional and scientific societies, philanthropic and corporate foundations, aquaria, botanical gardens, museums, science centers, after-school providers, and private industry, for example, play potentially significant roles in growing our Nation's STEM education pipeline and creating pathways to STEM. Each stakeholder brings a set of resources and expertise that are necessary for our Nation's STEM education systems to reach their full potential. The Federal Government works closely with these stakeholders to identify common areas of concern and to collaborate on strategies that leverage Federal and non-Federal assets, and this Strategic Plan creates an opportunity for more of these key partnerships. Efforts such as those described in BOX 1 illustrate the potential additive power of partnerships that amplify the contributions of the Federal Government and other stakeholders.

BOX 1. *100kin10* **Coalition**

Building on a recommendation of the President's Council of Advisors on Science and Technology (PCAST), President Obama in his 2011 State of the Union address called for a new effort to prepare 100,000 excellent STEM teachers over the next decade with strong teaching skills and deep content knowledge.

In response to that call to action, more than 150 organizations to date have joined together in a coalition called *100Kin10* to make over 100 measurable commitments to increasing the supply of excellent STEM teachers; hiring, developing, and retaining excellent STEM teachers; and building the 100Kin10 movement. With leadership from Carnegie Corporation of New York, the coalition and its partners have raised over $52.5 million from a broad range of foundations and philanthropists under a unique "funding marketplace" model where funders can choose from among a registry of high-quality proposals. The 100Kin10 coalition has also become a network with great potential for disseminating ideas, with leading-practice case studies, capacity-building workshops, and funds available for competitive research and collaboration grants.

---

4. Strategic Priorities and Coordination Strategies

This Strategic Plan provides a shared roadmap to ensure that the collective Federal investment of 14 agencies in STEM education has the best potential for substantial impact in key priority areas. In addition, the Strategic Plan has been developed after carefully considering the findings of CoSTEM *Portfolio* reports analyzing Federal investments in STEM. (See Appendix A for the *FY 2011 Federal Science, Technology, Engineering and Mathematics (STEM) Education Portfolio Highlights.)*

Strategic vision: A future where:

- The United States has a well-qualified and increasingly diverse STEM workforce able to lead innovation in STEM-related industries and to fulfill CoSTEM agency workforce needs;

- American students have access to excellent P-12, postsecondary, and informal STEM education and learning opportunities; and

- Federal STEM education programs are based on evidence and coordinated for maximum impact in priority areas.

Achieving this vision will require strong CoSTEM agency commitment to prioritization and coordination. The remainder of this plan sets forth five priority STEM education strategic investment areas and two goals for STEM education coordination.

4.1 Priority STEM Education Investment Areas

The priority areas identified by CoSTEM were selected on the basis that enhanced and coordinated Federal investments in these areas seem especially likely to accelerate progress toward the strategic vision outlined above. These areas reflect the needs of our Nation, align with priorities of both the Administration and Congress, and can draw effectively on Federal STEM-related assets. Through this Plan and Federal agencies' implementation efforts, the intention is to achieve significant, measurable impacts in five priority STEM education investment areas: (1) improve P-12 STEM instruction; (2) increase and sustain youth and public engagement in STEM; (3) improve undergraduate STEM education; (4) better serve groups historically underrepresented in STEM fields; and (5) design graduate education for today's STEM workforce.

Over the course of the implementation of the Strategic Plan, Federal agencies will strive to align their STEM education investments in support of those five priority investment areas. Through this effort, existing collaborations will be strengthened, and new collaborations will be undertaken. It is likely that additional areas of importance will emerge over time, and as progress is made on these priorities they may be adapted and focused. Also, it is important to note that these national goals are meant to provide overall direction and focus, and that Federal agencies will need to work together to achieve them. Progress on how Federal agencies are contributing will be assessed by developing metrics and milestones for the strategic objectives, introduced in section 5.

1. **Improve STEM Instruction.** *Prepare 100,000 excellent new K-12 STEM teachers by 2020, and support the existing STEM teacher workforce.*

Research shows that top-performing teachers make a dramatic difference in student achievement and suggests that for students who learn from these teachers year after year, achievement gaps narrow significantly.[33, 34, 35, 36, 37, 38] Every STEM student deserves an excellent teacher – one whose preparation has included appropriate opportunities to learn his or her subject and to participate in clinical experiences in schools. Every teacher and education leader deserves access to the preparation, on-going support, recognition, and collaboration opportunities needed for success. However, not all teachers receive the support they need, and science and mathematics teachers are among the most difficult to recruit and retain in K-12 schools.[39, 40]

To increase the number of excellent K-12 STEM teachers, CoSTEM agencies will undertake increased coordination among STEM-teacher preparation, professional development, support, and recognition efforts within existing and proposed programs. Continued research on teacher learning and STEM teacher development is essential to guiding these implementation and coordination efforts. Appropriate connections to local and state policy, standards, and assessments will be encouraged to ensure that the Federal investments are designed for local impact and implementation.

2. **Increase and sustain youth and public engagement in STEM.** *Support a 50 percent increase in the number of U.S. youth who have an effective, authentic STEM experience each year prior to completing high school.*

Progress for the Nation in STEM depends not only on a STEM workforce, but on a public that understands the role of STEM in addressing societal issues and is prepared to use STEM knowledge in personal and professional settings. For many Americans, both students and adults, opportunities for STEM learning occur through effective engagement across classroom and out-of-school settings, as each year tens of millions—of all ages and backgrounds — engage with science in informal settings ranging from pre-school programs to museums, to national parks, as well as cyberspace, all of which provide potential venues for advancing science literacy. This engagement[41] is critical to the learning process and to selection and persistence in STEM careers. Through effective engagement, learners have the chance to develop interest in and positive attitudes toward STEM topics, as well as improved perception of their ability to participate in STEM, all of which can lead to improved understanding and proficiency.[42, 43] Exciting trends today related to STEM engagement include digital badges that encourage students to earn progressive recognition in the form of online and mobile distinctions for acquiring cumulative skills, the Maker Movement, low-cost tools that allow students to design and make just about anything, and "games for learning." Additionally, citizen science activities can advance both science and learning.

By "authentic STEM experience," CoSTEM means a designed experience inside or outside of school in which learners engage directly in doing STEM. This broad designation covers a range of commonly referenced notions, from "hands-on" science, to problem-based learning, to inquiry. Part of the challenge in implementing steps toward this goal will be defining and studying whether experiences that are purported to be "authentic" have measurable impact on student motivation, persistence, and learning.

Agency implementation of this priority area will involve the development of coordinated programs inside and outside of school, platforms, and infrastructure to provide desired audiences with STEM experiences enabled through government assets, including but not limited to materials, facilities, and skilled STEM professionals. CoSTEM agencies - many of which employ skilled scientists and engineers, develop advanced instruments and facilities, and conduct research in STEM - currently play a key role in both formal and informal education engagement efforts. CoSTEM agencies also generate a wealth of information and data electronically available to the public, which can enable exploration of science or engineering subjects as part of classroom learning to encourage general interest, or as citizen science. Collaboration among all agencies will be critical for better understanding the current baseline of engagement activities and their outcomes and furthering a focused and impactful government investment in engagement.

3. **Enhance STEM experience of undergraduate students.** *Graduate one million additional students with degrees in STEM fields over the next 10 years.*

Poor retention rates among undergraduate STEM majors in U.S. institutions remain a major concern. Only 43 percent of students entering as a STEM major in a four-year public college or university graduate with a STEM degree. Worse, only about 14 percent of community college students who declare a STEM major on entry are still in a STEM field at the time of their last enrollment.[44]

Economic projections suggest the need for as many as one million additional STEM professionals over the next decade above current graduation rates.[45, 46, 47] That includes STEM majors at a variety of skill and knowledge levels from community college graduates, to quality STEM teachers, to scientists and engineers with advanced degrees. Outside of STEM careers, other fields increasingly require employees to have a good foundation in STEM disciplines. In addition, a scientifically and computationally literate citizenry is increasingly necessary to critically evaluate personal and societal issues that increasingly rely on science and technology underpinnings.

Through internship, scholarship, and fellowship programs, research experience opportunities, initiatives to create and test innovative instructional approaches and materials, faculty professional development, and research on STEM learning, Federal agencies can create strong emphasis on improving undergraduate STEM education. In addition, efforts to broaden participation in STEM will be critical to reaching the number of needed graduates, and the resulting diversity will enhance innovation in STEM fields.

4. **Better serve groups historically underrepresented in STEM fields.** *Increase the number of underrepresented minorities that graduate college with STEM degrees in the next 10 years and improve women's participation in areas of STEM where they are significantly underrepresented.*

Groups underrepresented in STEM fields include Hispanics and Latinos, African Americans, American Indians, Alaska Natives, Native Hawaiians and Pacific Islanders, the economically disadvantaged, people with disabilities, and women and girls. According to the most recent census, underrepresented minority groups (URGs) make up approximately 28 percent of the U.S. population but are experiencing the greatest population growth—with an 11 percent increase for African Americans and a 37 percent boost for Hispanics between 2000 and 2009.[48] Minority populations are expected to increase to 54 percent of the U.S. population by 2050[49, 50] and student diversity is reflected in this demographic change. Although the student population is becoming increasingly diverse, URGs remain underrepresented in STEM education and careers. Moreover, women, who comprise 48 percent of the U.S. workforce, have remained below parity and make up only 24 percent of STEM professionals. Underrepresentation of these groups in STEM fields begins early and persists across the P-12, post-secondary, and STEM-workforce spectrum.

To make progress in this priority area, CoSTEM agencies will work with relevant stakeholder communities, including faculty, administrators, and students from Minority-Serving Institutions (MSIs) given that approximately 10 percent of the funds invested by Federal agencies on this priority are focused on these institutions. Agencies will consider emphasizing education at critical transition points from P-12 to postsecondary education and from postsecondary education to the STEM workforce, when students from groups traditionally underrepresented in STEM often drop out of the STEM pipeline. As part of this focus, agencies will work to create more common definitions and consistent categorization of programs that serve underrepresented groups as either a focus or an emphasis. This may also include improving access to and increasing coordination across, programs for Minority-Serving Institutions.

5. **Design graduate education for tomorrow's STEM workforce.** *Provide graduate-trained STEM professionals with basic and applied research expertise, options to acquire specialized skills in areas of national importance and mission agency's needs, and ancillary skills needed for success in a broad range of careers.*

The National Center for Science and Engineering Statistics projects that between 2010 and 2020, 2.6 million jobs will require an advanced degree.[51] Filling these positions will be critical to the Nation's economic prosperity and global competitiveness. Recent reports by the Council of Graduate Schools, the NIH, the National Research Council, and professional societies[52] have called for shifts in approaches to STEM graduate education. The Federal Government supports graduate students primarily through research assistantships on individual investigator awards, which generally involves student participation in the research activities of the principal investigator. Mechanisms such as fellowships and traineeships also provide significant opportunities for graduate students, and increasingly there is a need to prepare students not only for academic research positions, but also for career options in the private and government sectors.

As an initial focus under this strategic investment priority area, CoSTEM agencies will coordinate to improve access to, and efficacy of, government-funded graduate fellowships. Over time, CoSTEM agencies may also consider addressing improvements to a broader range of Federal approaches to graduate student support.

Federally funded graduate fellowships recognize students with high potential in STEM research and innovation. These fellowships provide an opportunity for students to pursue unique educational opportunities and research across disciplinary boundaries, often with significant autonomy. Importantly, fellowships are designed to provide support directly to individual students to pursue their research interests. Thus, seeking a better way to leverage and coordinate fellowships provides a promising starting point in promoting and encouraging new opportunities to prepare the science and engineering workforce of tomorrow. There is growing evidence that opportunities for professional development of students to learn a broader range of skills that are important in STEM fields (e.g., communication) and to participate in applied work on authentic problems and challenges of government and the private sector are important components of graduate education.

Furthermore, tomorrow's STEM workforce will need to include effective change makers and entrepreneurs in business, public service, civil society, and academia. Some universities are encouraging students to set and meet more ambitious goals for their research, education, and service; giving students greater autonomy earlier in their career; connecting students to real-world problems at a regional, national, and global level; and involving students in the design of university curricula, research initiatives, and collaborations with external partners. The CoSTEM agencies can continue to benefit from understanding the effectiveness of innovative practices that universities are incorporating in their graduate education to promote those practices in graduate education more broadly.

4.2 STEM Education Coordination Approaches

Coordination of Federal investments in STEM education, including mission-agency STEM-workforce training and education, is central to the success of this Strategic Plan is. The two coordination approaches are (1) building new models for leveraging assets and expertise; and (2) identifying, using, and sharing evidence-based approaches. The intent is to establish a coordinated, coherent portfolio of STEM education investments across the Federal Government so efforts and assets are deployed effectively and efficiently, for greatest potential impact. By designating initial lead and collaborating agencies within certain priority areas, the Strategic Plan will encourage a more deliberate focus among new and existing efforts and create new synergies among programs and agencies. Likewise, many agencies already are committed to the use of evidence and rigorous evaluation in budget, management, and policy decisions, and to building the internal capacity to do so. That can also be supported and enhanced in a number of ways, including by creating and using common metrics, evidence guidelines, and evaluation practices; by developing complementary program goals; and by creating a shared understanding of evidence-based STEM education practices.

1. **Build new models for leveraging assets and expertise.** *Implement a concept of lead and collaborating agencies in priority areas to leverage capabilities across agencies to ensure the most significant impact of Federal STEM education investments.*

Efficient and effective use of funding and resources remains a priority for Federal investments, including those in STEM education. Efficiency across agencies can be created through inter-agency coordination and collaboration. Efficiency within agencies will occur through alignment of investments with capabilities, roles, and missions. An emphasis on investments designed to be complementary, that leverage available resources, will lead to a strategic portfolio intended for maximum impact.

To achieve this leveraging, lead agencies have been identified in several priority areas for this Strategic Plan. All agencies will employ a range of strategies to bring coherence and efficiency into STEM education programs with similar goals, including aligning programmatic goals, building common infrastructure, developing joint solicitations or memoranda of understanding, consolidating programs, and using new funding strategies, such as performance partnerships that encourage agency collaborations.[53]

Lead agencies will be responsible for convening other CoSTEM agencies, for helping to facilitate review and revision with collaborating agencies, and for tracking progress toward achieving Administration priorities. In each priority area, all CoSTEM agencies will be invited to participate. The following have been designated as the initial lead agencies for CoSTEM implementation:

- Department of Education (ED), which will assume a convening role to initiate planning around the priority area of "Improve STEM Instruction." The early focus of these discussions will be around how to best mobilize Federal assets and resources toward the improvement of STEM teacher preparation and continuing teacher professional development; how to connect these efforts with state and local policies and contexts; and how to leverage partnerships for improvement.

- National Science Foundation (NSF), which will play a convening role around the priority area of "Enhance STEM Experience of Undergraduate Students," including assisting with improving the delivery of undergraduate STEM education through evidence-based reforms. NSF will also work alongside NIH, USDA and other CoSTEM agencies on improving graduate fellowships. Efforts will include sharing detailed information about agency and disciplinary STEM workforce needs at the post-baccalaureate and postdoctoral levels, and mapping existing Federal resources and assets as a basis for designing strategic collaborations and synergies.

- Smithsonian Institution, which will focus its convening efforts on the priority goal of "Increase and Sustain Youth and Public Engagement in STEM." As part of the effort, Smithsonian will work with NSF, ED, the other CoSTEM agencies including NASA, NOAA, USDA, NIH, DOI, and other science partners best understand the agencies' unique expertise and resources for engaging learners with science, and will explore existing and potential approaches to improving infrastructure and access.

Designation as a lead agency does not narrowly define an agency's role in STEM education. It does not mean that NSF, for instance, will abandon its efforts in areas other than the improvement of undergraduate STEM education, or that no other agency will have any role in undergraduate education. It means instead, that NSF may request additional funding and resources to support an increased role initially in leading the improvement of undergraduate STEM education. Similarly, while ED is the initial lead for P-12 instruction, it will also play a strong role in supporting engagement activities and building bridges between in-school and out-of school learning to increase the effectiveness of both. The other CoSTEM agencies will be key collaborators, working with the lead agencies to find ways to build on their existing investments in STEM education, and leverage the passion and expertise of their staff and other STEM professionals who will continue to provide access to STEM content and Federal assets that can be used in formal and informal learning environments.

Following on the work that will be done within the initial convenings around the priority areas, CoSTEM will create implementation subcommittees for the STEM education investment priority areas and interagency coordination approaches, as appropriate. The Implementation Subcommittees would be responsible for tasks including:

- Reviewing roadmaps and implementation plans prepared by initial lead and collaborating agencies;
- Tracking implementation of the education investment priority areas within the Strategic Plan;
- Developing and monitoring metrics for progress, assessing whether efficiencies and improved impact are achieved, and recommending adjustments to the implementation process; and,
- Developing a framework and process for more coordinated Federal STEM education budget planning.

2. **Build and use evidence-based approaches**. *Conduct STEM education research and evaluation to build evidence about promising practices and program effectiveness, use across agencies, and share with the public to improve the impact of the Federal STEM education investment.*

There is a strong interest across the government in using evidence in policy and management decisions in all arenas, including STEM education, to increase the impact and efficiency of Federal programs. Given the important role of data and evaluations, Federal STEM education investments will increase use of research studies and program evaluation to identify and test practices and approaches for their effectiveness, thus building the evidence base for improving Federal STEM education investments. As appropriate, programs will rest on or scale up evidence-based practices. Findings about effectiveness will be shared, along with research-based understandings, and where indications of success are strong, agencies will be encouraged to act upon them. Where evidence is only suggestive of promise, but there are other rationales for making investments of a particular type, programs and investments will be designed so that an evidence base is built as the program is implemented.

Where appropriate, CoSTEM Implementation Subcommittees, together with lead and collaborating agency staff, will determine the most suitable approaches to synthesize evidence-based practices and processes for accumulating data, evaluations, information, and metrics and will consider collaborative evaluation approaches within and across agencies.

CoSTEM has developed a set of design principles (see Appendix B) that represent initial frameworks for Federal investments to be successful based on current evidence and best practices. As appropriate, these will be updated over time as new research and findings emerge. In addition, the impact and utility of information produced by STEM education investments will be considered in evidence-based grant-making strategies (e.g., pay for success and tiered-evidence grants). There is also work that is available to begin to identify such practices within the What Works Clearinghouse (http://ies.ed.gov/ncee/wwc), including reviews of studies of interventions and practice guides that identify opportunities for additional development and validation work.

5. Implementing the Federal STEM Education 5-Year Strategic Plan

For each of the five priority STEM education investment areas, a set of strategic objectives has been identified as the focus for initial implementation of the Strategic Plan. These objectives have been developed with several considerations: they strive to be specific, so that progress and impact can be measured; they are meant to align with the strengths and assets of the designated lead agency and to allow for the significant role of collaborating agencies; and they represent areas where there is a clear responsibility for involvement of the Federal Government, with the realization that Federal investment will play only a part in achieving the intended impact.

Implementation of the Strategic Plan will require commitment of the Administration, agency leadership, and legislative leaders in allocating the resources for the collaboration, coordination, and evaluation that will be necessary to realize these goals. Implementation also will require oversight and the development of more detailed roadmaps by lead and collaborating agencies over the coming months. CoSTEM will assist in this process.

5.1 Implementation of STEM Education Priority Investment Areas

In order to make progress on the priority STEM education investment areas, lead and collaborating agencies will determine which features of current Federal investments could be adapted, modified, or leveraged in support of the coordinated activities of the lead agencies, and how. This collaborative process will ultimately determine how agencies can best create infrastructure in priority areas and how significant strengths and assets at collaborating agencies can be made available in conjunction with these efforts. This will require coordination of agency priorities and goals, statutory requirements, authorizing legislation, and available expertise and resources. Close collaboration among lead and collaborating agencies will be needed to maximize the opportunities to continue to meet any mission-specific needs.

Preliminary implementation roadmaps are provided below for each of the five priority STEM education investment areas that, where appropriate, may help guide future budget planning and requests, mechanisms for investment, communication and outreach with stakeholder communities, and re-assessment of evaluation plans and practices within each of the priority areas. The ability of agencies to implement the strategic objectives and to create common metrics will depend on agency capacity and will require financial commitments to ensure adequate capacity both at lead agencies, for design and oversight of programs, and at collaborating agencies, for asset identification and coordination functions. The implementation of the Strategic Plan will include a process for periodic examination of overall progress on all strategic objectives, with the expectation that some may be revised, and others will be added.

Implementation Roadmap: **Improve STEM Instruction**

IMPACT STATEMENT: Prepare 100,000 excellent new K-12 STEM teachers by 2020, and support the existing STEM teacher workforce.

The education and development of teachers of STEM involves their pre-service preparation and also continuing professional development and learning opportunities over the course of their careers. The retention of teachers includes strategies that can involve school- and district-based resources and professional supports, meaningful community-based initiatives, and exciting opportunities to learn STEM and participate in authentic STEM research. Excellent STEM teachers know how to provide effective instruction and to inspire interest in these subjects so that their students understand STEM concepts and skills, and the Federal Government has a strategic role in providing learning and growth opportunities for the STEM teaching workforce.

Background: Of all of the activities that occur within the formal and informal education systems, the interactions among teachers, learners, and the content is the primary determinant of student success in grades K-12. Research shows that top-performing teachers can make dramatic differences in student achievement and suggests that the impact of assigning students to top-performing teachers each year can significantly narrow achievement gaps.[54, 55, 56, 57] Thus, the need to focus on improving STEM teaching is clear.

The scale of K-12 STEM education is large. There are more than 3.6 million full-time teachers in the Nation's elementary and secondary schools.[58] Of these, about 500,000 teach at least one course in either mathematics or science at the middle- or high-school levels.[59] Nationwide, there are about 2,800 universities and colleges that prepare teachers. These institutions are accredited by states or national organizations to certify teachers. Additionally, there are "alternative" programs to prepare teachers. According to the National Center for Alternative Certification, there are 136 state-defined alternate routes to teacher certification and about 600 alternate route programs that are implementing these options.[60] Policies for teacher credentialing and continuing licensure are developed at the state level, as are standards for K-12 STEM curricula.

The need for more STEM teachers is well documented,[61] but the recruitment and retention of STEM teachers is challenging. States routinely list a lack of teachers in mathematics and the sciences as one of their most pressing labor shortages.[62] The pre-service preparation of STEM teachers, activities that occur largely within individual institutions of higher education and are governed by state policies, does not compare favorably to the preparation of teachers in countries whose students show high levels of achievement in STEM.[63] Additionally, there is continued debate on the specific STEM content that teachers need to be effective.[64,65] The emerging field of discipline-based educational research (DBER) documents the difficulties of teaching undergraduate STEM courses for understanding and, along with PCAST's *Engage to Excel* report, promotes widespread use of evidence-based instructional practices – including in undergraduate courses where future teachers are learning STEM content.[66] And, as technologies that can support and improve classroom learning are readily available, teachers need to have opportunities to learn and use these exciting and powerful tools.

Once teachers complete their initial preparation, local districts and schools plan and pay for the majority of their professional development activities, and much of this professional development is focused on general classroom and pedagogical issues rather than subject-specific topics. Federal investment in teacher professional development is, conservatively, only 9% of the total Federal STEM investment (Appendix A). The quality of this professional development ranges considerably, and research is limited about its impact on teachers' effectiveness in enabling student learning. [67, 68, 69] In addition, CoSTEM agencies often offer a wide range of learning opportunities intended to inspire STEM teachers. In light of recent findings that access to learning opportunities and school supports are critical in keeping STEM teachers in classrooms,[70] it is important that the professional development made available to teachers is connected to their day-to-day work, related to state standards (including the Common Core State Standards in Mathematics and the Next Generation Science Standards) and to the ongoing professional credentials that are required in states.

Federal investments across lead and collaborating agencies alone cannot possibly reach all STEM teachers, or ensure that the offerings provided are compatible with the policies and practices of the states, districts, and schools where teachers are located. Thus, the focus for the Federal investments in teacher education, both pre-service and in-service, needs to be on building well-tested, replicable models that can be scaled or adapted across teacher education institutions, and at the level of states and districts for implementation, and providing infrastructure that can allow for scaling. These models can then be aligned with local policies and processes in ways that support teachers in their schools and classrooms. In addition, developing partnerships with local entities, local and regional science facilities that can support high quality STEM content, and the private sector will be especially critical.

In his 2011 State of the Union address, President Obama called for a new effort to prepare 100,000 STEM teachers over the next decade with strong teaching skills and deep content knowledge. The President's call built on key conclusions of the President's Council of Advisors on Science and Technology (PCAST) - that teachers need to have enough content knowledge to link STEM to compelling real-world issues, model the process of scientific investigation, effectively address student misconceptions, and help their students learn to reason and solve problems like mathematicians, scientists and engineers.

Significant Related Efforts: The *FY 2011 Portfolio* report indicates that, for FY 2011, about 10 percent ($315 million) of Federal STEM Education funding was dedicated to investments reporting to have a primary objective of supporting pre-service and in-service STEM educators. Federal agencies funded an additional $925 million in investments that include improvement of STEM teacher education as a secondary goal. Of the investments that primarily focus on improving teacher education, 78 percent supported the professional development of in-service educators, and the remainder supported both pre-service and in-service educators. The majority of the dollars came from the NSF and ED, through large programs such as ED and NSF's Mathematics and Science Partnership (MSP) programs, Teacher Loan Forgiveness, and the Robert Noyce Scholarship Program.

Strategic objectives. ED will serve initially as the lead agency for the goal of improving K-12 STEM instruction. All other CoSTEM agencies will serve as collaborating agencies. Collectively, Federal agencies will focus investments that would contribute to achieving the President's goal of preparing 100,000 new excellent STEM teachers, with strategies such as:

- Identify, develop, test, and support effective teacher preparation efforts that encourage teachers' use of evidence-based practices that provide students with rich STEM learning opportunities; and
- Increase the number and quality of authentic STEM experiences[71] for pre- and in-service P-12 teachers participating in federally supported internship, fellowship, and scholarship programs

Table of potential actions/outcomes/metrics:

Identify, develop, test, and support effective teacher preparation efforts that encourage teachers' use of evidence-based practices that provide students with rich STEM learning opportunities		
Actions	**Outcomes**	**Metrics/Milestones**
Near-term (years 1-2)		
Identify and assess Federal investments that incentivize (a) the recruitment or training of excellent K-12 STEM teachers to work in high-need schools and (b) the retention of those effective STEM teachers whose students show annual growth in STEM learning.	Better understanding of Federal investments that incentivize (a) the recruitment or training of excellent P-12 STEM teachers to high-need schools and (b) the retention of those effective STEM teachers whose students show annual growth in STEM learning.	Federally funded STEM teacher education initiatives that: (a) provide data on recruitment or training of excellent STEM teachers in high-need schools; (b) provide data on the retention of effective STEM teachers in high-need schools; (c) include data about teacher education practices, both pre-service and in-service education and experiences; (d) relate participation in these programs to student outcomes; and (e) as applicable, hold state educational agencies and local educational agencies accountable for implementing evidence-based practices that lead to (1) recruitment of excellent P-12 STEM teachers to high need schools and (2) retention of effective teachers whose students show annual growth in STEM learning.

Mid-term (years 3-4)		
Use the evidence base amassed to develop guidance for effective Federal programs that incentivize (a) the recruitment or training of excellent P-12 STEM teachers to high-need schools and (b) the retention of those effective STEM teachers whose students show annual growth in STEM learning.	Collecting research findings about the nature of Federal programs that incentivize (a) the recruitment or training of excellent P-12 STEM teachers to high-need schools and (b) the retention of those effective STEM teachers whose students show annual growth in STEM learning. As appropriate, Federal program investments implement guidance to increase the percentage of excellent P-12 STEM teachers (a) recruited or trained to teach in high-need schools and (b) retained when their students show annual growth in STEM learning.	Federally funded STEM teacher education programs including pre-service and in-service education and experiences, that: (1) have incorporated feedback to monitor the use of evidence-based practices and (2) monitor the evidence to be aware of changes that could impact programs.
Long-term (4+ years):		
Federal funding initiatives are designed to include: (1) recruitment and retention incentives; (2) evidence-based practices; and (3) the collection of necessary data to document the impact of teachers' participation in the Federal programs on (a) the increased percentage of excellent P-12 STEM teachers recruited to high-need schools, and (b) the increased percentage of effective STEM teachers retained whose students show annual growth in STEM learning.	Regular updates on the status of Federally-funded teacher programs that incentivize (a) the recruitment or training of excellent P-12 STEM teachers to high-need schools and (b) the retention of those effective STEM teachers whose students show annual growth in STEM learning. Maintain the advances obtained and institutionalize accountability for program outcomes.	Federally funded STEM teacher education programs in each agency, which offer pre-service and in-service education and experiences focused on recruitment and retention, will monitor the quality of the teacher education programs involved and assess the impact of the investment on improving (a) the recruitment of excellent P-12 STEM teachers to high-need schools and (b) the retention of those teachers whose students show annual growth in STEM learning.

Increase the number and quality of authentic STEM experiences[72] for pre- and in-service P-12 teachers participating in Federally supported internship, fellowship, and scholarship (IFS) programs.

Actions	Outcomes	Metrics/Milestones
Near-term (years 1-2)		
Identify and assess Federal STEM investments in teacher IFS, including: size, scope, structure, methods, status of assessment and evaluation activities and characteristics of included STEM experiences. Initiate plans for linking existing resources with new infrastructures for reaching wide audiences.	Collect information about Federal investments in teacher IFS including recommended additional information beyond that which was available. A baseline of relevant programs will provide the basis upon which assessment criteria can be developed. These criteria will be used to compare/contrast practices to develop guidance.	Agencies with teacher IFS that: (1) provide participant data on IFS; (2) include data about teacher education practices; (3) hold organizations accountable for implementing evidence-based practices; and (4) are capable of connecting participation in authentic STEM experiences with desired pupil outcomes.
Mid-term (years 3-4)		
Use the evidence base to develop guidance for effective program design of scholarships to support teacher candidates.	Investments implement guidance to increase effectiveness.	Agencies with teacher IFS that: (1) have incorporated feedback to monitor the use of evidence-based practices and (2) monitor the evidence to be aware of changes that could impact programs.
Long-term:		
Federal IFS funding initiatives are designed to include: (1) evidence-based practices; (2) grade-appropriate authentic STEM experiences; and (3) collection of necessary data to document the connection between teachers' participation and desired student outcomes.	Biennial report on the status of Federally-funded teacher IFS programs. Maintain the advances obtained and institutionalize accountability for program outcomes.	All Federally funded teacher IFS programs monitor the quality of the teacher education programs involved.

Implementation Roadmap: **Increase and Sustain Youth and Public Engagement in STEM**

IMPACT STATEMENT: Support a 50 percent increase in the number of U.S. youth who have an effective, authentic STEM experience each year prior to completing high school.

Engagement is part of the larger process of learning. It is the critical component for capturing the learner's interest and involvement, and for inspiring further development of knowledge and understanding. To ensure a STEM-skilled workforce of the future, the Federal Government has considerable assets that can engage youth so that the pathway through their education leads to the challenging STEM-related careers of tomorrow, and to a "culture of STEM" in the public.

Background: Given the importance of STEM engagement at a national level, a common understanding of the terminology is necessary. As defined in *The 2011 Federal STEM Education Portfolio*, investments that focus on engagement are designed to increase learners' involvement and interest in STEM, inform their view of STEM's value in their lives, or positively influence the perception of their ability to participate in STEM. The scope of STEM engagement is vast, and includes investments in a wide range of areas, such as development of learning materials; programs at museums, science centers, or parks; games, simulations, and virtual environments; "Citizen Science" initiatives; public talks, and educational broadcast programming. Thus, there are many avenues available to the Federal Government for reaching learners, and many possibilities for tailoring content for STEM-learning audiences.

Research indicates that instructional approaches or learning opportunities that engage students actively increase skill acquisition and information retention, encourage more positive attitudes toward STEM disciplines, and strengthen retention of students in STEM majors.[73, 74] For example, learning theory and empirical evidence about how people learn[75] suggest that STEM experiences that engage learners in "active learning" improve retention of information and critical thinking skills.[76] Furthermore, research studies in STEM education support this positive relationship between STEM engagement experiences and student achievement. For example, one study demonstrated that a 2- to 3-week university Summer Science Academy program, which provided students in grades 6-12 with field trips, laboratory experimentation, and other authentic, science-related, learning opportunities nonexistent in the students' schools, produced significant, long-term outcomes on student achievement and attitudes toward science.[77] This growing body of research literature on effective STEM engagement and evidence-based practices can guide Federal Government investments.

Additionally, agencies are already exploring or supporting a number of promising new and emerging strategies in the area of youth STEM engagement and have started to generate great momentum within grassroots networks and crowd-driven virtual platforms. For instance, the Maker Education Initiative, which grew out of a network of self–directed STEM enthusiasts who share their creations with students and the general public through Maker Faires across the world, recently announced its intention to create a Maker Corps of student maker leaders. These youth, embedded in informal learning institutions across the country, will provide hands-on learning opportunities to interested students, exciting their desire to be, as the President has said, "makers of things, not just consumers." Additional new opportunities for engagement include the digital badge movement bolstered by NASA[78] as well as the MacArthur and Mozilla Foundations. Additionally, crowd-sourced collection of data through "citizen science" activities like "Backyard Biofuels," provide an opportunity to evaluate and scale exciting new avenues for inspiring students in STEM fields while building an evidence base for novel activities.

In the years ahead, the Smithsonian Institution will convene agencies to help improve the reach of informal STEM learning opportunities by ensuring that materials are aligned with what students are learning in the classroom as appropriate, and that exciting, emerging science is accessible to students through innovative means. Priority will be placed on effective, authentic STEM experiences. The Smithsonian will work with the CoSTEM agencies such as NASA, USDA, NIH, DOI, NOAA, and other science partners to harness their unique expertise and resources to enable the use of promising materials and curricula, on-line resources, and effective delivery and dissemination mechanisms by more learners both inside and outside the classroom.

Significant Related Efforts: Over half of the Federal Government's STEM education investments identified in *The 2011 Federal STEM Education Portfolio* provided engagement as the primary or secondary objective (156 of 233 programs). Funding for investments with the primary objective of engagement totaled $164 million. Of these investments, about half (approximately 44 percent) are categorized as large-scale programs. The budget of each investment was under $30 million.

CoSTEM agencies are heavily involved in engagement. For example, USDA houses the Cooperative Extension System through land-grant universities and includes a mission to bring research, education and extension, and applied research-based practices, to both youth and adults. This occurs through 4-H, a unique public-private partnership with USDA, extension land-grant universities and local governmental entities providing public support along with a private, non-profit national partner, the National 4-H Council. Together with private donors and foundations, these entities support the programmatic efforts for over 6 million youth participants and half a million volunteers, both youth and adults. Since 2010, 4-H's Science Leadership Academies have engaged 1,266 4-H professionals in training workshops with the ultimate goal of equipping more 4-H staff and volunteers with the ability to engage students in informal learning environments.

The Corporation for National and Community Service (CNCS) is launching a new STEM AmeriCorps, multi-year initiative to place hundreds of AmeriCorps members in nonprofits across the country to mobilize STEM professionals with the goal of inspiring young people to excel in STEM education.[79] The U.S. Department of Education also hosts a network of 21st Century Community Learning Centers to provide academic enrichment opportunity during non-school hours for children in high-poverty and low-performing schools. The Smithsonian sponsors a Youth Engagement through Science (YES!) effort, which connects local youth with Smithsonian collections, experts, and training to inspire them to pursue STEM careers.[80]

The Federal Government has a diverse range of assets to increase interest in STEM. These assets include, but are not limited to scientific data, technology, research and engineering facilities, natural environments, science-technology centers, engineers, technologists, and scientists. The agencies use these assets to provide online, place-based (e.g., research and engineering facilities, federally managed lands and waterways, and museums or visitor centers), and other experiential or hands-on learning opportunities primarily in informal learning environments for people of all ages. In addition, NSF provides development and research grants in the area of informal STEM education, while some CoSTEM agencies have legislative mandates to promote direct public participation in exploration, STEM research, and environmental stewardship. The private sector can help in these efforts, leveraging the passion and expertise of its science and technology trained workforce. For example, leading U.S. technology companies have launched an effort called US2020, an all-hands-on-deck effort to have many more STEM professionals mentor children from kindergarten through college.[81]

Alignment, coordination, and continued research on such experiences can support the development of multiple pathways to continued and deeper exploration of STEM learning in informal learning environments, connections between formal and informal learning settings,[82] and the impact of engagement experiences on outcomes such as student learning. Together, these efforts will move the STEM engagement enterprise forward while building evidence-based practices. In this rich and diverse context for providing engagement, the Smithsonian will assume an initial lead role for coordination and developing the infrastructure to allow the more efficient and open access[83] to these extraordinary Federal resources.

A key focus within the Engagement priority is on assessing the impact of engagement experiences with appropriate outcome measures and better understanding the connections between inspiration and excitement and persistence and achievement in STEM for school and careers.

Strategies: With the Smithsonian playing an initial lead coordination role, Federal agencies will focus investments on three main strategies.

- Federal investments in engagement that draw explicitly on the scientific, technological, and engineering assets (e.g., facilities, scientific and engineering staff, instruments, data, and federally-managed public lands and waterways) of the Federal Government, where feasible, to provide authentic experiences;
- Federal engagement investments that support the integration of STEM into existing school readiness and after-school programs with significant local, regional, or national reach; and
- Federal engagement investments that contribute to improving empirical understanding of how engagement in authentic STEM experiences relates to improved student learning or interest outcomes.

Table of potential actions/outcomes/metrics:

Federal investments in engagement that draw explicitly on the scientific and engineering assets (e.g., facilities, scientific and engineering staff, instruments, data, and Federally managed public lands and waterways) of the Federal Government to provide authentic experiences.		
Actions	**Outcomes**	**Metrics/Milestones**
Near-term (years 1-2)		
Identify scientific and engineering assets that are being effectively leveraged in existing investments on engagement. Build additional infrastructure to provide access to Federal engagement assets.	Identify agencies and assets that have evidence-based models of best practices in engagement.	Number of collaborations that draw on STEM assets with evidence of impact for engagement.
Mid-term (years 3-4)		
Establish new approaches to make mission agency assets available through new networks and other agency collaborations to wider groups of youth.	Identification of opportunities to form connections between engagement assets and programs.	Number of youth that have significant interactions with Federal science agency assets.
Long-term:		
Revise solicitations beyond Federal science mission agencies to tie education research and development to the assets of the mission agencies.	A research and development infrastructure to assess and track impact of engagement experiences.	Increase instruments and assessment techniques that provide valid, reliable assessment of engagement and identification of relevant outcome measures.

Federal engagement investments that support the integration of STEM into existing school readiness and after-school programs with national reach.		
Actions	**Outcomes**	**Metrics/Milestones**
Near-term (years 1-2)		
Identify and begin collaborations among a set of programs for school readiness and after school with national reach, including exploration of possibilities with Head Start, 21st Century Community Learning Centers, and 4-H.	Convening to identify model programs and mechanisms to facilitate collaborations.	Number of projects that are being leveraged for national reach.
Mid-term (years 3-4)		
Launch partnerships among agencies and programs to develop curricula, professional development, & implementation.	Development of engagement-related materials and models for partnerships are supported by agencies.	Number of projects that have links to programs with national reach.
Long-term:		
Initiate longitudinal study to examine the relationship between introducing STEM engagement experiences into national programs and student outcomes.	Report on research findings about relationship between authentic STEM experiences and engagement-related outcomes in national programs.	Number of learners reached through national programs that have embedded evidence-based STEM engagement experiences.

Federal engagement investments that contribute to improving empirical understanding of how engagement in authentic STEM experiences relates to improved student learning or interest outcomes.		
Actions	**Outcomes**	**Metrics**
Near-term (years 1-2)		
Determine which investments are positioned to collect data on authentic STEM experiences and how they affect student learning, interest, involvement, and/or motivation (engagement) outcomes. Build knowledge base related to youth engagement and criteria to record number of authentic STEM experiences achieved through investments."	Identify availability and quality of data and methodologies to gather data on authentic STEM experiences.	Number of projects from their investments that are participating in the building of the knowledge base of authentic STEM experiences with mechanisms for relating to student engagement.
Mid-term (years 3-4)		
Continue data gathering and analysis efforts, and when appropriate, the development of new metrics and methodologies.	As appropriate, implementation of a data coordination infrastructure to enable research across sites to better understand the relationship between authentic STEM experiences and engagement related outcomes.	Number of projects from their investments that are participating in the building of data coordination infrastructure.
Long-term:		
Encourage Federal engagement initiatives to participate in the data coordination infrastructure.	As appropriate, report on research findings about relationship between authentic STEM experiences and engagement related outcomes.	Have many more Federally funded authentic STEM engagement projects are part of the data coordination infrastructure.

Implementation Roadmap: Enhance STEM Experience of Undergraduate Students

IMPACT STATEMENT: Graduate one million additional students in STEM fields over the next 10 years.

A number of economic and labor analyses suggest that if the United States is to maintain its global preeminence in the STEM fields and benefit from the social, economic, and national security advantages that come with such preeminence, then it must produce approximately one million more STEM professionals than are projected to graduate over the next decade.[84, 85, 86] To meet this goal, the United States institutions of higher education will need to increase the number of students who receive undergraduate STEM degrees by about 34 percent over current rates by 2020. To meet this need, the Federal Government has made preparing one million more STEM graduates the central focus of the Cross Agency Priority (CAP) goal on STEM education, a multi-agency effort to align agency activities and accountability around a clear priorities.[87] The Federal Government, however, is just one of a number of stakeholders needed to reach the goal of graduating one million additional students with STEM degrees, and so attaining it will involve public-private partnerships as well as a Federal focus.

Background: Students with STEM degrees are needed at a variety of skill- and knowledge-levels—from community college graduates trained in information technologies or those involved in the production of advanced manufacturing, materials and energy, to high-quality STEM teachers, to scientists and engineers working in research labs. Besides STEM career areas, other fields increasingly require employees to have a good foundation in STEM disciplines. In addition, it is increasingly important to prepare a scientifically and computationally literate public able to critically evaluate societal issues as diverse as climate change, application of medical technologies, or alternative energy sources. Also in the competitive global work environment there are a number of other competencies that are considered useful[88] such as curiosity, creativity, tolerance of ambiguity, resilience in the face of setbacks, and abilities to work effectively with people who have different perspectives, priorities, or intellectual approaches. Experiences within the STEM education system may provide opportunities for building these competencies.

Poor retention rates in undergraduate STEM majors remain a major concern and a critical factor in achieving the one million more graduates goal. Only 43 percent of students entering a STEM major in a four-year public college or university graduate with a STEM degree. Worse, about 14 percent of community college students who declare a STEM major on entry are still in that field at the time of their last enrollment.[89] PCAST concluded in its recent report, *Engage to Excel: Producing One Million Additional College Graduates with Degrees in Science, Technology, Engineering, and Mathematics*, that retaining more STEM majors is the lowest-cost, most efficient policy option to provide the STEM professionals that the Nation needs.

As noted in the PCAST report, as well as in the Federal CAP goal, increasing the retention of STEM majors to 50 percent would generate approximately three-quarters of the targeted one million additional STEM graduates over the next decade by increasing the annual number of students with bachelor or associate degrees in STEM fields by approximately 75,000. Furthermore, such an increase appears feasible.

Research has uncovered a variety of evidence-based best practices to recruit, engage, and retain STEM majors.[90] These practices include pedagogies, curricula, instruction materials, peer or mentors and other academic and cultural supports, resources, and tools to engage students in the classroom and support their learning. Also, there is evidence that co-curricular activities—such as learning communities and summer "bridge" programs—that provide academic support, and community experiences that encourage students to adopt STEM majors, can have a positive impact on student retention.[91] Application of such practices at both two-and four-year schools, coupled articulation agreements that ensure compatibility of both discipline content and pedagogical practice, are promising strategies to increase significantly the number and diversity of talented STEM majors entering the workforce and graduate education.

BOX 2. Department of Defense (DoD) and NSF: Awards to Stimulate and Support Undergraduate Research Experiences (ASSURE)

Research experiences for undergraduates are important authentic STEM experiences that provide a means for students to gain first-hand appreciation of the knowledge-production process. The Awards to Stimulate and Support Undergraduate Research Experiences (ASSURE) program is a collaboration between DoD and NSF to increase student engagement in STEM. DoD's ASSURE is executed through NSF's Research Experiences for Undergraduates (REU) program to support undergraduate research in DoD-relevant disciplines at NSF-sponsored college and university locations. ASSURE seeks to increase the number of high-quality undergraduate science and engineering majors who ultimately decide to pursue advanced degrees in those fields.

With a well-defined common focus, students gain access to faculty and/or other research mentors and relevant facilities to complete research projects. The research projects can be based in a single discipline or academic department, or in interdisciplinary or multi-department research opportunities with a strong STEM focus, providing a range of possible research experiences for undergraduate students.

Significant Related Efforts: The Federal Government has a number of investments that can be leveraged to improve undergraduate STEM education. Some investments already include collaboration across agencies (See BOX 2). Historically, the majority of funding from investments that at least partially target undergraduate students has the primary goal of graduating students with STEM degrees. The other primary objectives, with at least $100 million dedicated to support undergraduate students, were in the categories of education R&D, institutional capacity building, learning, and STEM careers. An additional substantial portion of funding has been dedicated to programs that provide fellowships, scholarships, and/or internships. In addition, many agencies have programs in place to offer research experiences to undergraduates.

Furthermore, students should be empowered to play a more active role in their own educations through student-led innovation. For instance, the MIT Clean Energy Prize is a student-led, student-focused clean-energy innovation and entrepreneurship prize that has helped train hundreds of students from more than 60 universities around the country with the skills necessary to be clean-energy entrepreneurs. More than 30 companies have been launched as a result of the competition since the prize was founded in 2008.[92] Student-led innovation programs currently allow agencies to directly encourage this community of 20 million technically capable, creative, and energetic innovators to use their skills to work on real-world problems. Engaged and empowered students can increase the ability of universities to organize research, coursework, and experiential learning and to become more engaged universities.

In addition, as outlined in the CAP goal on one million more STEM graduates, the private sector is also taking leadership in addressing challenges of undergraduate STEM education.[93] For example, through such initiatives as the National Undergraduate STEM Partnership, leading universities, industry groups, the Business Higher Education Forum, the Association of American Universities, the Association of Public and Land-Grant Universities, the American Council on Education, the National Defense Industrial Association, and others have committed to improving undergraduate STEM education, particularly in the first two years of college. Private industry also has sought ways to partner with the Federal Government to encourage increased enrollment and retention in STEM fields. For example, through its Graduate 10K+ Initiative, the NSF has partnered with Intel and GE in making grants to institutions whose projects aim to improve retention of undergraduates in engineering and computer science. This effort is funded with $10 million in private-sector donations.[94]

Strategic objectives: To contribute to the goal of producing one million more STEM graduates in the next 10 years, and building upon the CAP goal in this area, the Federal Government, working closely with partners and with NSF will pursue the following strategies:

- Identify and broaden implementation of evidence-based instructional practices and innovations to improve undergraduate learning and retention in STEM and develop a national architecture to improve empirical understanding of how these changes relate to key student outcomes;
- Improve support of STEM education at 2-year colleges and create bridges between 2- and 4-year postsecondary institutions;
- Support and incentivize the development of university-industry partnerships, and partnerships with federally supported entities, to provide relevant and authentic STEM learning and research experiences for undergraduate students, particularly in their first two years; and
- Address the problem of excessively high failure rates in introductory mathematics courses at the undergraduate level to open pathways to more advanced STEM courses.

In these strategies, agencies will work to incorporate a focus on increasing the participation of women and underrepresented minorities and identifying and supporting innovation in higher education.

Table of potential actions/outcomes/metrics:

Identify and broaden implementation of evidence-based instructional practices and innovations to improve undergraduate learning and retention in STEM and develop national architecture to improve empirical understanding of how these changes relate to key student outcomes		
Actions	**Outcomes**	**Metric or Milestone**
Near term (years 1-2)		
As appropriate, establish a cross agency mechanism to obtain data on instructional practices at institutions.	Revisions to solicitations, syntheses of evidence-based practices, and measurement of the use of evidence-based instructional practices introduced in programs, as appropriate.	Number of programs that introduce a focus on evidence-based instructional practices.
Mid-term (years 3-4)		
Agencies partner with stakeholders to develop and implement faculty professional development to use evidence-based practices/design innovation	New partnerships among agencies, professional societies, and higher education to focus on development and use of evidence-based practices.	Improved retention in STEM majors in a selected set of universities participating in partnerships.
Long-term:		
Stakeholders participate in national effort to validate and scale-up evidence on relationship of practices to student outcomes such as retention.	Infrastructure for sharing national data about use of evidence-based practices and relationship to student outcomes has been developed.	National findings available about the relationship between instructional and program characteristics in undergraduate STEM and student outcomes.

Improve support of STEM education at 2-year colleges and create bridges between 2- and 4-year postsecondary institutions.

Actions	Outcomes	Metric or Milestone
Near-term (years 1-2)		
Gather information about STEM programs and challenges at two-year colleges.	Identification of issues in STEM at two-year colleges.	Better understanding of STEM education at two-year colleges and next steps for agencies
Mid-term (years 3-4)		
Establish network of two-year colleges and 4-year institutions committed to STEM excellence, in partnerships as possible.	Stronger linkages and better articulation agreements between two year and four year colleges and universities and science assets among Federal agencies.	Have many institutions of higher education committed to partnering with Federal agencies to improve 2-to-4 year transitions and STEM opportunities.
Long-term:		
Federal agencies reconfigure or enhance programs as appropriate to support effective partnerships.	Programs and STEM outcomes improve in 2-year colleges.	Number of two-year colleges with effective STEM programs substantially increases.

Support the development of university-industry partnerships, and partnerships with Federally supported entities, to provide relevant and authentic STEM learning and research experiences for undergraduate students, particularly in their first two years.

Actions	Outcomes	Metric or Milestone
Near-term (years 1-2)		
Identify existing Federal STEM internship and other authentic research opportunities, and assemble available evidence about impact.	A more detailed baseline description of the Federal portfolio.	Federal agencies can better report the number of internships and other STEM research opportunities available to undergraduates, and any available evidence about their impact.
Mid-term (years 3-4)		
Convening of government, higher education, and private sector to share best practices and develop incentives for research and internship experiences for students in the first two years of undergraduate education as called for in the PCAST engage to excel report and the CAP goal.	More commitments from industry and government to increase the availably of internships and other authentic research opportunities for students in the first two years.	Federal programs increase focus on research experiences in the first two years and on assessing impact.

Long-term:		
Federal agencies lead the development of a streamlined application process and potentially establish a database of available research experiences.	Industry and government combine resources for reaching more students.	Increased diversity in the students who participate in undergraduate research experiences in the first two years.

Address the problem of excessively high failure rates in introductory mathematics courses at the undergraduate level, to open pathways to more advanced STEM courses.		
Actions	**Outcomes**	**Metric or Milestone**
Near-term (years 1-2)		
As feasible, NSF and ED prepare a joint solicitation to improve K-16 mathematics education. As feasible, accounting of mathematics funding in agency portfolios and using that to inform the solicitation.	Joint work between NSF and ED in this key area.	Programs that incorporate the expertise or assets of more than one agency to improve passing rates of introductory mathematics courses.
Mid-term (years 3-4)		
Convening on introductory college mathematics, including government and key stakeholders to highlight promising practices, potential partnerships, and key challenges.	Spread of information about effective practices and new partnerships for scaling and identification of challenges and changes	A cohort of colleges and universities that have significant challenges in introductory mathematics performance have initiated changes in programs and practices.
Long-term:		
National data collection/surveys redesigned to track the transitions of students from high school to undergraduate courses in mathematics.	New databases to allow for monitoring national progress on the mathematics issues.	A cohort of universities and colleges can demonstrate substantial improvement on introductory mathematics performance.

Implementation Roadmap: **Better Serve Groups Historically Underrepresented in STEM Fields**

IMPACT STATEMENT: Increase the number of underrepresented minorities that graduate college with STEM degrees in the next 10 years and improve women's participation in areas of STEM where they are significantly underrepresented.

Members of racial and ethnic minority groups are projected to become the majority of America's population in the next 30 years. Currently, however, they account for only 28 percent of STEM workers.[95] Women make up nearly half of the total workforce, but they constitute only 24 percent of STEM jobholders.[96] To meet growing demands from industry for more STEM-qualified employees continued and improved support for the development of talent from all groups is essential. This will require significant improvements in broadening the participation of underrepresented groups and women in STEM fields.

Background: The underrepresentation of certain groups in STEM fields begins early and persists across the STEM school and workforce spectrum. There are three complementary reasons why participation of underrepresented minorities, lower-income children, and women in STEM is critical to sustaining our capacity to conduct research and innovate: our sources for the future STEM workforce are currently not sufficient; the demographics of our domestic population are shifting dramatically; and diversity of ideas and perspectives in STEM is a strength that benefits both diverse groups and the Nation as a whole.[97, 98] Intentional and consistent efforts are needed along the STEM trajectory in order to increase the number of individuals from groups traditionally underrepresented in STEM that graduate and are well-prepared with STEM degrees, because members of these groups leave STEM majors at higher rates than others. These students deserve special attention and must be a deliberate part of any national strategy because they offer an expanding pool of untapped talents and are a large underutilized source of potential STEM professionals.

While the overall increase in college enrollment is encouraging, the National Academy of Sciences (NAS) report, *Expanding Underrepresented Minority Participation* [87] cited that STEM degree completion rates for underrepresented minorities (URMs) were still lower than that of White and Asian students. Moreover, URMs attending two-year colleges are less likely to major in STEM subjects and earn STEM degrees. In 2009-10, African Americans earned 8.6 percent of all science and engineering bachelor's degrees, including only five percent of all bachelor's degrees in engineering[99] Hispanics earned seven percent of all bachelor's degrees in engineering and computer and information sciences.[100] While women who pursue STEM degrees earn higher proportions of bachelor's degrees in medical and social sciences, they earn lower proportions of bachelor's degrees in engineering and computer and information sciences. In 2009-10, 18 percent of all bachelor's degrees in engineering and computer and information sciences were awarded to women.[101] Data also show that students with disabilities are less likely to major in a STEM field than students without disabilities.[102]

The disparities in access to STEM courses and programs that exist in higher education are also evident at the K-12 level. Data from ED's Office for Civil Rights' Civil Rights Data Collection (CDRC) reveal disparities in access to high-level mathematics and science courses in high school.[103] Notably, of the high schools serving the most Hispanic and African-American students, less than one-third offer calculus, and only 40 percent offer physics. Students with limited English proficiency make up six percent of the high school population, but account for 15 percent of the students for whom algebra is the highest-level mathematics course taken by the final year of their high school career. Only 19 percent of all girls taking at least one Advanced Placement (AP) course were enrolled in AP mathematics compared to 26 percent of all boys taking at least one AP course. Of the African-American or Hispanic students enrolled in AP mathematics, only 39 percent are girls. Girls also represented only 46 percent of the students enrolled in physics. According to CRDC data, students with disabilities represent four percent of students taking Algebra II, chemistry, and physics, 0.1 percent of students taking calculus, and eight percent of students taking biology.

Significant Related Efforts: Promoting equitable participation of people seeking STEM degrees and careers in STEM is a priority of a significant number of Federal programs. In FY 2011, $616 million in STEM education investments identified their primary focus as supporting underrepresented groups. More than 20 Federal STEM education investments, including the 1890 scholars programs at USDA, targeted historically underrepresented groups by supporting Minority-Serving Institutions (MSI), including Historically Black Colleges and Universities (HBCU), Hispanic-Serving Institutions (HSI), Alaska Native-Serving Institutions, Native Hawaiian-Serving Institutions, Insular areas institutions, and Tribal Colleges and Universities. In the past, the largest amount of funding was directed to HSIs, mostly from ED's Developing Hispanic-Serving Institutions STEM and Articulation Program, with additional funds at supporting agencies.

Of the Federal investments that are focused on improving participation of underrepresented groups in STEM (see Appendix A), the majority include support for undergraduate students or institutions (70 percent), while 20 percent include support for K-12 students. Also, for those investments that are focused on broadening participation in STEM, about 5 percent of funding supported research to better understand the science of broadening participation.

Although there has been some progress and limited success in broadening participation of underrepresented groups in STEM, much more can be done to attract and retain these groups. Some of the programs that have shown promise in their efforts in broadening participation include the Educational Partnership with Minority Institutions funded by NOAA (see BOX 3), the LSAMP program at NSF, and non-Federal programs such as the Meyerhoff Scholarship program at University of Maryland Baltimore County, the successful Bayer Scholars program of Duquesne University, and the revamped Computer Science program at Harvey Mudd College that has increased the percentage of women computer science majors from single digits to 40 percent in just a few years.[104] These programs have provided strong mentoring support, academic development and enhancements, and social and professional networks, and have helped students not only to complete their baccalaureate degrees but also to pursue graduate degrees that strengthen their abilities to obtain STEM jobs.

Moreover, there is a growing body of research demonstrating that teaching students in STEM classes that their intelligence is not a fixed trait, but instead is something that can be developed with effort and help from others (i.e., teaching them a "growth mindset"), can significantly improve academic outcomes, in some cases reducing achievement gaps among minorities and women by 30-50 percent.[105, 106]

BOX 3. NOAA Educational Partnership Program with Minority Serving Institutions

The National Oceanic and Atmospheric Administration (NOAA)'s mission is to understand and predict changes in the Earth's environment and conserve and manage coastal and marine resources to meet our Nation's economic, social, and environmental needs. To maintain a workforce with the scientific skills and technical capacity to meet and advance this mission, the agency requires a diverse pool of candidates that reflect the American population. To that end, in 2001 NOAA established the Educational Partnership Program (EPP) with Minority Serving Institutions (MSI) to: (1) increase the number of post-secondary students, particularly from underrepresented communities, who are educated in STEM disciplines that directly support NOAA's mission; (2) contribute to NOAA's mission by strengthening and building capacity in NOAA science and management areas primarily at MSIs; and, (3) leverage NOAA funds to strengthen the education, training and research capacity at the EPP Cooperative Science Centers (http://epp.noaa.gov).

Each of the four Cooperative Science Centers (CSCs), selected via a competitive grants program, is led by an MSI that collaborates with at least six colleges or universities. In addition to the CSCs, the EPP has a Graduate Sciences Program, an Undergraduate Scholarship Program, and an Environmental Entrepreneurship Program. Since its inception, the program has supported more than 1,100 students obtain undergraduate and graduate degrees in fields related to NOAA's mission; 119 of those students obtained PhDs. To date, 19 percent of science professionals from underrepresented communities who have been hired by NOAA had been supported by EPP.

Additionally, the Federal Government also has engaged in a number of collaborative efforts alongside private and non-profit and academic sector partners with the primary or secondary intention of increasing opportunity and access for students who are typically underrepresented in STEM. For example, the Equal Futures Partnership, a multi-country initiative launched by then-Secretary of State Hillary Clinton and other world leaders in September 2012, features commitments from a diverse set of corporate partners seeking to open doors to high-quality educational and career opportunities for women in STEM disciplines.

Strategic objectives: In order to contribute to the goal of improving STEM graduation success rates for underrepresented minorities and women, Federal agencies will work to implement three strategies:

- Be more responsive to the rapidly changing demographics and issues for particular groups and particular STEM fields through investments in broadening the participation of groups traditionally underrepresented in STEM;
- Focus investments on developing and testing strategies for improving STEM preparation for higher education for students from groups underrepresented in STEM; and
- Invest in efforts to create campus climates that are effective in improving success for students from underrepresented groups through mentorship, technical assistance, and other innovative practices.

Table of potential actions/outcomes/metrics:

Be more responsive to the rapidly changing demographics and issues for particular groups and particular STEM fields through investments in broadening the STEM participation of groups traditionally underrepresented in STEM.		
Actions	**Outcomes**	**Metric or Milestone**
Near-term (years 1-2)		
Agencies and stakeholders identify specific issues and concerns and encourage new ideas, from communities, and the field. Specifically focus on challenges associated with increased access to AP courses by military children, underrepresented minorities; individuals with disabilities, and on special initiatives to encourage women in computer and physical sciences and engineering. Where possible, increase participation of targeted underrepresented minorities, lower-income children, individuals with disabilities, military children, and women in internships at Federal research and engineering facilities.	Where appropriate, convening with MSIs and other stakeholders, together with Federal agencies, to develop common understandings of key challenges.	Proposed ideas for improving impact of Federal portfolio of investment in broadening participation across agencies.
Mid-term (years 3-4)		
Develop key partnerships among agencies and external organizations to design new courses or redesign existing ones to make them relevant to a growing number of URMs in STEM as well as make them competitive for entrance into college.	Have several agencies partner and develop mechanisms that encourage participation of URMs in taking Advanced Placement and other courses, (including on-line) which will allow them to gain college credits in various STEM courses.	Increased support for more relevant courses for URMs will be enhanced, thereby increasing opportunities for URMs in STEM.
Long-term:		
National databases and systems allow for more detailed monitoring of progress by group and STEM field.	Agencies and professional organizations are able to monitor progress of involvement of particular underrepresented groups in particular fields.	Increased participation of groups underrepresented in particular STEM fields is evidenced through national data sources.

Focus investments on developing and testing strategies for improving STEM preparation for higher education for students from groups underrepresented in STEM.

Actions	Outcomes	Metric or Milestone
Near-term (years 1-2)		
Review current models of supporting MSIs to identify gaps and complementary models in the Federal program portfolio and create logical pathways and linkages.	Where appropriate, conduct studies on the various models used by Federal agencies and design of new programs to fill gaps.	New programs will address gaps in the models and will increase the ability of the agencies to support more students from URMs and women, while complementary programs will offer more opportunities and access to students.
Mid-term (years 3-4)		
Synthesis and analysis of evidence-based practices emerging from research and development of more common metrics to assess the impacts of broadening participation using these analyses.	A set of guidelines or measures will be available to "test-drive" on Federal programs that support broadening participation.	Federal investments are coordinated to build knowledge base about effective strategies, and use common metrics to assess program impacts.
Long-term:		
Agencies will work with academia, industry, and education organizations to identify mechanisms that should be more widely implemented.	Graduation rates in STEM fields in both two-year and other higher education institutions for students from all groups traditionally underrepresented in STEM are increasing in key STEM fields	Clearer understanding of how effective approaches can be adapted for a variety of institutions.

Invest in efforts to create campus climates that are effective in improving success for students from underrepresented groups through mentorship, technical assistance, and other innovative practices.		

Actions	Outcomes	Metric or Milestone
Near-term (years 1-2)		
Coordinate efforts to ensure that students belonging to URGs are not excluded in the campus environment by engaging STEM education experts.	In FY2013, Federal agencies will develop a document that will serve as a guideline for a campus climate conducive to all students, particularly those belonging to URGs and will be disseminated to grantees, stakeholders and other relevant organizations.	Campus activities, policies and practices will reflect awareness of and agreement with the guidelines.
ED's Office for Civil Rights (OCR) will continue to update a STEM resources webpage it recently created to make clear to schools, students, and parents that the civil rights laws enforced by OCR apply to STEM courses and programs.	Enhanced understanding among stakeholders of civil rights steps needed to improve awareness of and compliance with regard to access to Title IX within STEM programs and contexts.	In FY 2013 will establish a baseline for providing guidance and tools to improve compliance.
Mid-term (years 3-4)		
Federal Government encourages higher education institutions that can be model demonstration sites for effective STEM campus climates.	Increased visibility for institutions that are effective in STEM-friendly campus climates for underrepresented groups, within a year of implementation.	Programs increasingly directly address campus climate.
Long-term:		
Higher education institutions that are not MSIs but that have significant numbers of students from underrepresented groups agree to adapt and implement policies and practices to change campus climate for more effective STEM education.	A gathering of institutions sponsored by professional higher education organizations to highlight the challenges of changes in campus climate.	Increased investment in research in the science of broadening participation as it relates to campus climate issues.

As appropriate, CoSTEM will determine appropriate mechanisms for launching a national conversation among MSI leaders and others to engage the broad stakeholder community in determining appropriate steps for launching the implementation of these strategic objectives, including potentially designating a lead agency for the effort.

Implementation Roadmap: **Design Graduate Education for Tomorrow's STEM Workforce**

IMPACT STATEMENT: Provide graduate-level trained STEM professionals with basic and applied research expertise, options to acquire specialized skills in areas of national importance and mission agency's needs, and ancillary skills needed for success in a broad range of careers.

Background: The preparation of graduate students in science and engineering contributes to global competitiveness, producing the highly skilled workers of the future and supporting the research needed for a knowledge-based economy. It is projected that between 2010 and 2020, 2.6 million jobs will require an advance degree, including non-STEM and professional degrees.[107] Given this need, it is critical that the training of today's scientists and engineers, at all levels, but particularly specialized advanced education, take into account the speed of technological innovation, the nature and practice of science and engineering, innovation as an economic driver, and the challenging issues facing society.

Recent reports by the Council of Graduate Schools (CGS), National Research Council, and the NIH all have made recommendations for how to better prepare graduate students. Their recommendations include strengthening professional development and deepening employer-university engagement. Those recommendations also are reflected in a recent PCAST report[108] that underscores the need to provide incentives for industry to invest in research and university-industry partnerships and to change educational programs to prepare graduates for a broad range of careers. Given that approximately 50 percent of doctorate holders work in fields outside of academia,[109] it is critical that the training and preparation students receive enables them to succeed in the workforce.

According to *Science and Engineering Indicators* 2012,[110] there were about 440,000 full time graduate students in U.S. institutions in 2009. Of these about 81,000 or 18 percent were supported by Federal funds. Of those supported by Federal funds, over 70 percent were on supported as research assistants on research grants, about 10 percent on fellowships, 10 percent on traineeships, and the remaining on Federally funded teaching assistants and other mechanisms. These funding mechanisms are critical in developing the STEM workforce and advancing U.S. research and innovation.

CoSTEM will initially focus on coordination to improve the access to, and efficiency of, graduate fellowships in STEM. In this regard, NSF will coordinate with agencies such as NIH (biomedical), NASA (aerospace), EPA (environmental), etc., as well as other agencies to understand specific needs and support applied research with regard to particular STEM fields and to help ensure that students have access to specialized training in order to prepare the Nation's STEM workforce. Roles and responsibilities may be defined, for instance through Memoranda of Understanding established between NSF and other CoSTEM agencies where appropriate, to help ensure their interests and STEM workforce needs are addressed. Over time, CoSTEM agencies may also consider addressing improvements to a broader range of Federal approaches to graduate education.

Significant Related Efforts:

The Graduate Education Modernization (GEM) informal working group, which includes DOE, NIH and NSF, has noted that a critical issue confronting U.S. competitiveness is the alignment of competencies required for success in the broader STEM professional workforce with the preparation provided by graduate education and training. A desired outcome of a broad GEM effort is an expanded collaborative effort between academic institutions and Federal funding agencies to help ensure that the graduate education enterprise provides enhanced opportunities to prepare students for a wide range of career paths, without compromising preparation for the academic research enterprise that currently relies heavily on graduate students to advance knowledge and innovation. For example, NIH expects all recipients of the Ruth L. Kirchstein National Research Service Award institutional research training grants to provide trainees with professional development skills and career guidance.[111] More recently, NIH's the National Institute of General Medical Sciences has identified some useful ways to implement the actions in their Strategic Plan for Biomedical and Behavioral Research Training, such as encouraging exposure to multiple career path options for students and recognizing that well-prepared graduates must have a deep knowledge in a specific field and be conversant in related fields.[112]

The informal Interagency Workgroup for STEM Graduate Fellowships (IWGF), which has included the graduate fellowship directors from all the CoSTEM agencies, seeks to strengthen the ongoing and future diverse pools of highly trained scientists and engineers in the United States. The IWGF goals are to (1) discuss common issues among federally funded graduate fellowship programs and (2) create more efficient mechanisms across all stages of fellowships management ranging from pre-award (the competition and review processes) to post-award (fellow's professional development, graduation rates, etc.).

Strategic Objectives for Federally Funded Fellowships: In order to contribute to the goal of improving coordination on and access to graduate fellowships in STEM, Federal agencies will focus on three strategies:

- Recognize and provide financial support to students of high potential for contributions in science and engineering fields, and for success in their ultimate careers;
- Provide opportunities for fellows' preparation in areas critical to the Nation and, in particular, critical to preparing the workforce needed to advance the missions of Federal agencies, including scholarship-for-service and STEM-focused pathways to Federal service programs throughout the Federal research and engineering enterprise; and
- Continue and enhance mechanisms that evaluate the impact of fellowships to inform future Federal investments.

Table of potential actions/outcomes/metrics:

Recognize and provide financial support to students of high potential for contributions in science and engineering fields, and for success in their ultimate careers.		
Actions	**Outcomes**	**Metric or Milestone**
Near-term (years 1-2)		
Develop a coordinated Federal approach to fellowships that increases efficiency and effectiveness for agencies and applicants, including "one stop shopping" features.	Implementation of coordinated approach to fellowships that increases efficiency and effectiveness for agencies and applicants.	Increased number of opportunities in areas of national need offered to increased number of fellows.
Mid-term (years 3-4)		
Design of a comprehensive approach to fellowships that increase efficiency and effectiveness for agencies and applicants.	Established process for identifying areas of national need and developing opportunities in those areas.	Process identified with steps for implementation.
Long-term:		
Coordinated Federal approach to fellowships through cooperative effort of Federal agencies.	Efficient and effective Federal approach to providing fellowships for developing highly skilled S&E workforce, and preparing students in areas of national need.	Evaluation results that determine impact of Federal investments in fellowships and inform future investments.

Provide targeted opportunities for fellows' preparation in areas critical to the Nation and, in particular, critical to preparing the workforce needed to advance the missions of Federal agencies, including exploring scholarship-for-service and STEM-focused pathways to federal service programs throughout the Federal research and engineering enterprise.		
Actions	**Outcomes**	**Metric or Milestone**
Near-term (years 1-2)		
Develop mechanisms to increase the Fellows receiving opportunities for preparation in areas of national need, including, where appropriate, establishing interagency MOUs or other mechanisms of coordination between NSF and the other CoSTEM agencies .	Established process for identifying areas of national need and developing opportunities in those areas. □Evidence of quality preparation	Evidence of quality preparation in areas of national need as assessed against goals of mission agency. Numbers of STEM fellowship program graduates serving in federal workforce. Progress on meeting national STEM workforce needs.

Mid-term (years 3-4)

As appropriate, explore the enhancement of fellowship programs to create expedited pathways into public service for STEM professionals. Continued identification of opportunities in areas of national need.	Established infrastructure to offer Fellows development of opportunities in areas of national need	Increased number of opportunities in areas of national need offered to increased number of fellows.

Long-term:

Management of Federal comprehensive approach to fellowships through cooperative effort of Federal agencies.	Highly efficient and effective Federal approach to providing fellowships for developing highly skilled STEM workforce, and preparing students in areas of national need.	Evaluation results that describe impacts of Federal investments in fellowships and inform future investments.

Continue and enhance mechanisms that evaluate the impact of fellowships to inform future Federal investments.

Actions	Outcomes	Metric or Milestone
Near-term (years 1-2)		
Refine evaluation mechanisms to inform the improvement Federal fellowships for increased impact.	Where appropriate, longitudinal tracking system designed to provide useful information about impact of graduate experiences on preparation of STEM professionals.	Design of tracking system is accepted across Federal agencies.
Mid-term (years 3-4)		
Continued refinement and implementation of evaluation mechanism.	Implementation of evaluation mechanism.	Initial evaluation results on impact of fellowships toward developing highly skilled U.S. workforce, as well as workforce in areas of national need.
Long-term:		
Coordinated Federal approach to fellowships through cooperative effort of Federal agencies.	Highly efficient and effective Federal approach to providing fellowships for developing highly skilled S&E workforce, and preparing students in areas of national need.	Evaluation results that determine impact of Federal investments in fellowships and inform future investments.

5.2 Implementation of the Coordination Objectives

Coordination Objective 1: Build new models for leveraging assets and expertise.

Implement a concept of lead and collaborating agencies to leverage capabilities across agencies to ensure the most significant impact of Federal STEM education investments.

Funding for Federal STEM education efforts has been dispersed across multiple agencies in many different programs and initiatives – a model that makes use of specific agency strengths and assets. With this Strategic Plan, a new model for this portfolio of investments is proposed, with a concept of lead and collaborating agencies organized around each of the priority goals.

Background: With more than 200 STEM investments among 14 agencies there are certainly programs and investments that have complementary elements that can be coordinated, overlapping elements that can be considered for consolidation, and potential synergies that can be maximized for stronger impact. The new approaches outlined in this plan will enhance the potential for scale, the impact, and the quality of the Federal STEM education investment in each of the strategic priority areas.

Lead agencies will be responsible for convening CoSTEM agencies and setting a coherent and detailed agenda for their area of responsibility, assessing the collaborative assets of other agencies as part of the plan, moving the implementation plans forward in the respective goal areas, and ensuring that planning for evaluation and impact studies occurs. These activities will be coordinated with the collaborating agencies that bring unique assets, knowledge, and facilities to the table, and where appropriate, assisted by the CoSTEM Implementation Subcommittees. More efficient leveraging of funding and other resources across agencies is a critical component of building coherence in the investment portfolio. Alignment of investments will lead to increased efficiency within and across agencies as well as clarification of capabilities, roles, and missions.

Significant related efforts: There are a number of venues in which Federal agency staff with responsibility for STEM education investments regularly come together to collaborate and address major national issues. These activities will serve as the baseline foundation for the new model. For example, the Informal Science Education Interagency Working Group was recently created to develop increased coordination and collaboration among Federal agencies. Other groups also exist, such as the NSTC National Initiative for Cybersecurity Education (NICE), the Networking and Information Technology Research and Development (NITRD) program, the Environmental Education Taskforce, the Ocean Education Interagency working group, the U.S. Global Change Research Program Communication and Education Working Group, and the OMB evaluation group.

In addition, because agencies tend to interact with overlapping stakeholder communities, the annual meetings held by such professional societies as the American Mathematical Society, the Society for Research on Educational Effectiveness, the National Association for Research in Science Teaching, the American Education Research Association, the Council of Graduate Schools, the American Association for the Advancement of Science, National Science Teachers Association, American Geophysical Union, Association of Science and Technology Centers, the Council of State Science Supervisors, and others often include sessions that bring together experts from several agencies to discuss program offerings and connections. Some key Federal committees that are concerned with STEM education matters, such as the Committee on Equal Opportunities in Science and Engineering (CEOSE) and the National Board for Education Sciences have designated agency representatives or liaisons who participate in their meetings. Finally, there are a number of collaborative undertakings in STEM education that involve two or three agencies working together (e.g., the NSF-ED K-16 Mathematics Initiative and the ED-NASA Summer of Innovation and 21st Century Community Learning Centers work). These efforts, together with the efforts of CoSTEM, all provide a strong context for reaching the next levels of collaboration.

Strategies: To build and implement new models for leveraging and collaborating across agencies, Federal agencies will explore key strategies including:

- Collaborating in a new arrangement of lead and collaborating agencies to build implementation roadmaps in the goal areas;
- Designing new infrastructure, networks, and mechanisms to ensure that assets and resources available through Federal agencies are used and accessed widely for the improvement of STEM learning across the country, and reducing administrative barriers to collaboration; and
- Developing a framework to guide coordinated CoSTEM agency budget requests.

Table of potential actions/outcomes/metrics:

Collaborate in a new arrangement of lead and collaborating agencies to build implementation roadmaps in the goal areas.		
Actions	**Outcomes**	**Metric or Milestone**
Near-term (years 1-2)		
As appropriate lead and collaborating agencies will convene around relevant priority areas. Where appropriate, CoSTEM will create new Implementation subcommittees in the strategic priority areas.	Agreements among lead and collaborating agencies about detailed agenda and plans for progress in each strategic priority area.	Roadmap and detailed implementation timelines.
Mid-term (years 3-4)		
Track progress on implementation, collect evidence, and recommend mid-course corrections.	Evidence-based planning for appropriate adjustment and updating of strategic priorities and objectives in each area.	Implementation subcommittees could provide input to inform development of subsequent Administration STEM education budget requests.
Long-term:		
Recommend additional priority areas for longer-term improvement in STEM education. Take steps to assess effectiveness of this approach.	Continuous assessment of needed improvements in STEM education based on evidence in implementation in original five priority areas.	Concept papers developed or commissioned to assess needed emphases and priorities for long-term improvement of STEM education.

Design new infrastructure, networks, and mechanisms to ensure that assets and resources available through Federal agencies are used and accessed widely for the improvement of STEM learning across the country, and reduce administrative barriers to collaboration.

Actions	Outcomes	Metric or Milestone
Near-term (years 1-2)		
Assemble information about government mechanisms that relate to cross agency collaboration and recommend approaches for streamlining and simplifying to promote collaboration. Benchmark existing federal infrastructure, networks, etc.	More capacity to collaborate across agencies based on shared understanding of procedures and mechanisms.	Simplification of key processes such as development of MOUs and administrative clearances to encourage common procedures and collaborations.
Mid-term (years 3-4)/Long-term:		
Introduce innovative strategies for cross-agency capacity building and expertise development in STEM education.	Such innovations as shared personnel, common offices that span agencies, and joint funding opportunities.	An increased in joint funding opportunities that include more than two agency partners.

Develop a framework to guide coordinated CoSTEM agency budget requests.

Actions	Outcomes	Metric or Milestone
Near-term (years 1-2)		
As appropriate, convene CoSTEM agency representatives, including CFOs, to determine feasibility, timing and process for collaborative budget requests, potentially as early as the FY 2015 request.	Coordinating budget process for STEM education.	Production of cross-agency STEM education budget request framework.
Mid-term (years 3-4)/Long-term:		
Assess impact and effectiveness of the coordinated request framework and revise practices and approach accordingly.	Revised coordinated budget process for STEM education.	Continued collaboration on cross-agency coordinated STEM education budget request.

Coordination Objective 2: Build and use evidence based approaches.

Conduct rigorous STEM education research and evaluation to build evidence about promising practices and program effectiveness, use across agencies, and share with the public to improve the impact of the Federal STEM education investment.

An ever-growing body of education research and evaluation findings is improving our understanding of STEM learning and STEM instructional practices. These include a wide set of studies, including foundational work on the brain and cognition research on the critical role that "non-cognitive" factors play in aiding achievement in STEM fields, examinations of classroom design, implementation research on teacher education programs, and evaluations of large-scale education reform efforts. Findings from emerging research are more useful when they are synthesized and shared and when the studies build from existing evidence. In addition, gaps between fundamental research and studies to understand and validate promising practices persist.

Background: Federal STEM education programs are more likely to have a greater impact if they use, amass, and improve evidence of what works. There are substantial bodies of education research and evaluation studies that have influenced the general directions of improvement in STEM education. Basic findings about how people learn STEM and how to teach STEM subjects,[113, 114] for instance, are consistent with current emphases on active engagement with meaningful STEM learning settings. Similarly, there is a large body of work in the area of STEM education evaluation, at the level of projects, lines of activity (e.g., IES' evaluation of teacher professional development[115]), and full Federal programs/investments. Some evaluation methodologies also lead to insights about why, for whom, and under what conditions particular interventions or programs are effective or not. At the same time, there is substantial area expertise and experience within the STEM education staffs of the CoSTEM agencies that is deployed with the planning, revision, and implementation of any Federal investment. This professional knowledge has been refined over time and can serve to guide the effective planning and implementation of investment activities, including the design principles described in Appendix B.

Despite a rich body of work, more can be done to leverage the findings of research studies and evaluation efforts to strategically shape the direction of Federal investments. Given limited resources and the availability of some evidence of the most promising STEM education practices, it is important that those shaping government investments incorporate existing knowledge to make investments most likely to have the intended impact. At the same time, it is important to decide which research and evaluation questions can be addressed with new investment ideas, so the body of evidence continues to grow and our collective understanding is both improved and communicated.

Collaborative strategies for evaluating individual investments within and among agencies and themes that cut across several investments can improve the evaluation capacity of all agencies. Models such as the MIT Poverty Action Lab, or creative use of administrative data to undertake impact studies, as NIH and others have done,[116, 117] should be explored. In addition, agencies might consider developing cross-agency monitoring systems that will allow the collection, comparison, and analysis of uniform data across investments.

Significant Related Efforts: Syntheses of research are produced regularly by the National Research Council; recent examples, supported by NSF, have focused on K-8 science learning education,[118] learning science in informal environments,[119] STEM schools,[120] monitoring progress in STEM education,[121] discipline based education research,[122] and improving minority participation in STEM.[123] The most recent major synthesis of research on mathematics learning K-8 is the 2008 report of the President's National Math Advisory Panel.[124] ED has supported syntheses on many topics, including research on teacher preparation.[125] The Institute of Education Sciences (IES) has a searchable database that includes several funded synthesis studies; in addition, the IES-sponsored What Works Clearinghouse contains reviews of many interventions. Recent PCAST reports[126, 127] provide recommendations that are based on such research findings. Synthesis of evidence-based practices and research findings is an ongoing process, and additional syntheses are needed periodically as empirical findings continue to emerge. In addition, the National Research Council has reviewed agency education portfolios, providing valuable perspective on these efforts.[128, 129]

Agencies have also made strides in developing their evaluation capacity. For example, ED's IES has made rigorous, experimental, and quasi-experimental evaluation the norm in many education issue areas. Further, ED's Investing in Innovation program spearheaded a "tiered evidence" structure that tests new ideas, validates promising interventions, and scales up proven practices. Additionally, NSF has created a STEM Education Evaluation Working Group to improve evaluation of its own STEM education programs. Other agencies have been making continual improvements to their evaluation capacity during recent years and have completed agency-wide plans for evaluation of STEM education investments. As another example, DOI completed its first STEM education and employment strategic plan, which provides guidance that encourages a culture of evaluation to foster continual improvements and impact assessments. NOAA has embarked on an ambitious monitoring and evaluation effort to assess its education outcomes and impacts across programs. The purpose of this effort is to link outcome measures to program efforts and to evaluate the success of the agency in meeting its strategic goals.

Research and evaluation efforts also yield important methodologies and instruments that can be shared across government for use within research and evaluation efforts, such as the *Framework for Evaluating Impacts of Informal Science Education Projects*[130] produced through a multi-agency collaboration. The development of other instruments and tools is supported by programs including IES' Statistical and Research Methodology in Education grant program, and the NSF Promoting Research and Innovation in Methodologies for Evaluation (PRIME).

Agencies have already begun to develop shared strategies for generating evidence. ED and NSF have been collaborating on the development of a common evidence framework. The guidelines include suggestions for describing high-quality work within six education research study types: (1) foundational research; (2) early stage and exploratory research; (3) design and development projects; (4) efficacy studies; (5) effectiveness studies; and (6) scale-up studies.[131] The evidence guidelines are intended to be used by prospective grantees, Federal agency staff members, and merit reviewers.

The design principles developed by CoSTEM (see Appendix B) represent a synthesis of knowledge from research and practice about necessary conditions for effectiveness in each investment area. CoSTEM posted the design principles for public comment[132] and made adjustments based on the feedback received. Applying these principles to program design and implementation is intended to improve the prospects of accomplishing an investment's objectives and enhance cross-agency coordination.

Over the past several years, the Administration has laid a strong foundation and developed new tools that will help Federal agencies support evidence-based practices or strategies that will build useful knowledge of what works. For example, the Administration has encouraged the use of evidence-based grant program designs,[133] and identified approaches that can inform the development and continuous refinement of Federal grant-making. The guidance is based on the theory that program design and implementation strategies should build evidence about what works to increase Federal resources flowing to state, local, or other grantees using evidence-based practices and to identify more cost-effective practices. For example, a growing number of STEM education investments are implementing strategies such as tiered evidence frameworks.

Strategic objectives: To improve the exploration and sharing of evidence-based practices the Federal agencies will undertake the following strategic objectives:

- Support syntheses of existing research that can inform Federal investments in the STEM education priority areas;
- Improve and align evaluation and research strategies and expertise across Federal agencies, with lead agencies helping lead design of evaluation in their assigned goal areas; and
- Lower barriers to interagency collaboration by streamlining processes for interagency collaboration (e.g., MOUs, Interagency Agreements, shared staff, common instruments, etc.).

Table of potential actions/outcomes/metrics:

Support syntheses of existing research that addresses critical issues in STEM education priority areas.		
Actions	**Outcomes**	**Metric or Milestone**
Near-term (years 1-2)		
Identify critical research questions in each priority area working with stakeholder communities, and commission syntheses as appropriate.	Research synthesis and agenda in each priority area.	A series of research syntheses and agendas developed in collaboration with stakeholders.
Mid-term (years 3-4)		
As appropriate, agency investments are leveraged to address critical research questions.	Release requests for proposals to address critical research questions. NSF and ED will align some of their calls for research applications and proposals with critical research questions.	Funded research and evaluation studies and activities in the priority areas are assembled as a cross-agency portfolio and a monitoring system is introduced.

Long-term:		
Support symposia for each priority area with participation from stakeholders, researchers, practitioners, and Federal agency representatives.	Convening and consolidation of findings and sharing of ideas across multiple communities.	Analysis of literature and funded efforts in priority areas to examine nature of the knowledge base.

Improve and align evaluation and research strategies and expertise across Federal agencies.		
Actions	**Outcomes**	**Metric or Milestone**
Near-term (years 1-2)		
As appropriate, create a cross-agency STEM Education Evaluation working group to assess capacity and practices.	Agreement to inter-agency support of evaluation capacity in STEM education.	As appropriate, mechanisms for sharing STEM education evaluation capacity across agencies are established.
Mid-term (years 3-4)		
Agencies work together to implement a major STEM education investment evaluation for an investment in one of the priority areas that can provide evidence of impact, using current best practices and methodologies in evaluation.	A coordinated set of STEM education impact evaluations in the priority areas.	Agencies will collaborate in the implementation of impact evaluations in the four priority areas.
Long-term:		
Agencies develop monitoring systems to continue to document progress in priority areas.	New monitoring systems and new data accumulation.	Coordinated data systems and open access to information. Link to indicator and monitoring systems.

Lower barriers to interagency collaboration by streamlining processes for interagency collaboration (e.g., MOUs, Interagency Agreements, shared staff, common instruments, etc.).		
Actions	**Outcomes**	**Metric or Milestone**
Near-term (years 1-2)		
Agencies adapt Design Principles to fit context and shares across CoSTEM agencies.	Internal agency discussions about the process and considerations in investment design.	Revised Design Principles, incorporating agency details, made available across agencies.
Mid-term (years 3-4)		
Agencies introduce processes for documenting the use of Design Principles and evidence-based practices in the development of new STEM education solicitations or requests for proposals.	Internal change of culture in agency discussions about use of Design Principles and evidence in program planning.	New solicitations or requests for proposals incorporate in visible ways considerations of design principles and evidence.
Long-term:		
Cross-agency evaluation is planned to examine impacts of using design principles and evidence in development and revision of programs.	Documentation of how Design Principles and Evidence were used.	Process for tracking the impact of using Design Principles and evidence is in place.

5.3 Implementation Constraints

It is worth acknowledging external factors beyond the direct control of Federal agencies that may impede achievement of the goals in the Strategic Plan. Annual appropriations decisions by Congress will directly impact the flexibility that agencies have to coordinate and evaluate their efforts.

CoSTEM agencies have noted a number of potential constraints to successful implementation of this plan, including, among others :

- Overall agency budget fluctuations make long-term planning difficult.
- In some cases, authorizing language of some departments and agencies and sub-agencies limits them from targeting underrepresented groups.
- Challenges associated with collecting and sharing student data can limit STEM education program-evaluation strategies.
- For the most part, infrastructure and expertise for STEM education investments varies widely across departments and agencies.

6. Conclusion

Given that many jobs of the future will be STEM jobs, that our K-12 system is "middle of the pack" in international comparisons, and that progress on STEM education at multiple levels is critical to building a just and inclusive society, there is an urgent need to continue to improve STEM education in the United States. Much knowledge for how to make progress exists. Federal agencies have the collective capacity to make a difference. This Strategic Plan identifies goals, priorities, and a new framework and mechanisms for collaboration and program improvement. This Strategic Plan charts a course for sustained improvement and can help Federal investments in STEM education programs make a difference for many more students, educators, and members of the public. Implementation will require making hard choices, forging new partnerships, and focusing on outcomes. Adjustments along the way will be necessary but the Federal agencies will work together to make progress on this national priority.

Appendix A

The Federal Science, Technology, Engineering, and Mathematics (STEM) Education Portfolio: FY 2011 [134]

This Appendix details the results of CoSTEM inventory of Federal STEM education investments in FY 2011. The Fast-Track Action Committee on Federal Investment in STEM Education (FI-STEM) was chartered in April 2010 to develop the inventory process, analyze the inventory results, and draft a portfolio report and this update with the oversight of CoSTEM. The membership of the committee included representatives from the 12 Federal agencies that comprise CoSTEM. This appendix updates the original (FY 2010) inventory published in December 2011 as *The Federal Science, Technology, Engineering, and Mathematics (STEM) Education Portfolio* with an FY 2011 inventory. This inventory includes detailed information on STEM education investments in order to identify potential areas of synergy across and within agencies; support sharing of effective STEM education investment strategies and evaluation techniques across Federal agencies; increase awareness of STEM education investments within and across Federal agencies; and support development of a Federal STEM education 5-year strategic plan that is the main body of this report. The full FY 2011 inventory is Appendix Table A4.

The inventory process is also a catalyst for interagency collaborations related to improving investments in STEM education. By enhancing capabilities of the interactive database housing the inventory, CoSTEM anticipates that the inventory will serve as a tool to bring together staff working on similar efforts, facilitate their collaboration, and support their ability to implement program improvements.

Executive Summary of Appendix A

Over the past year, the NSTC's Committee on Science, Technology, Engineering, and Mathematics (STEM) Education (CoSTEM) completed an inventory of FY 2011 Federal investments in STEM education. This appendix is a report of the findings of that inventory which, together with the original FY 2010 inventory, informed development of the Strategic Plan that is the body of this report. Representatives from the 12 CoSTEM agencies developed the online survey used to collect the information. The FY 2011 survey was similar to the one used for the FY 2010 portfolio report.[135] Changes made to the FY 2010 inventory survey improved its clarity and the accuracy of the information collected.

The inventory data indicate that in FY 2011 the Federal Government expended $2.891 billion on more than 200 STEM-focused investments. Almost 80 percent of FY 2011 Federal STEM education funding was distributed through three Federal entities: the National Science Foundation (NSF) (40 percent); the Department of Health and Human Services (HHS) (20 percent); and ED (19 percent).

Additional findings from the analysis of the FY 2011 Federal inventory were useful for the development of the Strategic Plan, including the following:

- About one-third of Federal STEM education funding was spent on activities that target the specific workforce needs of Federal agencies surveyed, and the remaining two-thirds targeted broader STEM education activities.

- More than half of the funding (52 percent) was devoted to investments with a primary objective related to postsecondary STEM education (37 percent with a "STEM Degrees" objective; 15 percent with a "STEM Careers" objective).

- Thirty percent of funding went to investments that focused *primarily* on supporting underrepresented groups and that placed a competitive priority on supporting these groups, while an additional 60 percent of funding supported investments that *encourage* serving groups historically underrepresented in STEM fields, but do not make it a competitive priority.

- About 10 percent of Federal STEM education funding was dedicated to investments with the primary goal of supporting the preparation, education, or professional development of STEM teachers. An additional 33 percent of funding supported STEM teachers as a secondary objective.

The FY 2011 and FY 2010 portfolio reports provide a baseline CoSTEM can use to analyze funding trends and to identify areas potentially needing more investment.

The Federal STEM Portfolio: FY 2011

The *Federal Science, Technology, Engineering, and Mathematics (STEM) Education Portfolio: FY 2010* was the first of two interagency efforts to systematically survey and analyze data about Federal activities in science, technology, engineering, and mathematics (STEM) education. The inventory gathered information about how much the Federal Government spends on STEM education investments overall, and how many and what kinds of investments individual Federal agencies support. This examination of the FY 2011 Federal STEM education portfolio notes changes in investments from the previous year and examines the survey's impact on investment coordination and improvement.

CoSTEM chartered the NSTC Fast-Track Action Committee on Federal Investment in STEM Education (FI-STEM) in April 2010 to develop and administer the inventory. CoSTEM oversees the activities of FI-STEM, which includes members from the 12 Federal agencies represented on CoSTEM. FI-STEM met eight times between December 2011 and September 2012 to discuss outcomes of the previous inventory, revise criteria for the current inventory, update survey structure and definitions, determine collection and analysis methods, and develop a process for drafting this inventory report.

Criteria for Inclusion

For the inventory, detailed information was collected on Federally-funded STEM education investments in FY 2011 that had current and consistent funding at or above $300,000 annually (see Table A1 for the full criteria set). Only basic information such as budget, agency, investment name, evaluations conducted, focus on historically underrepresented groups, and funding mechanism was collected on STEM programs with budgets of less than $300,000, and/or that were funded by earmark or with American Recovery and Reinvestment Act funds. The decision to collect detailed information on only investments at or above $300,000 level was based on the judgment of FI-STEM members.

Table A1 details definitions, units of analysis, and other inclusion criteria used in the inventory. Other types of agency activities may contribute to STEM education and have been included in prior surveys (such as the *Report of the American Competitiveness Council*[136] and *Survey of DOD STEM Programs*[137]) but fall outside the scope of CoSTEM inventory. They include:

- research investments that fund science or engineering research and can support undergraduate or graduate students who assist in carrying out research if scientific research produced is the primary goal and measure of success;
- general education investments (such as Pell or Title I grants) that may include STEM as one of several education topics;
- volunteer activities by agency staff, such as classroom visits or judging STEM competitions, that do not require Federal funds;
- investments to promote public awareness of STEM education investments; and
- postdoctoral research awards or fellowships that support scientific research.

STEM education investments by the 12 CoSTEM agencies and the Nuclear Regulatory Commission (NRC) met criteria for inclusion in the inventory. The most common reasons that the activities were not included in the inventory were that the budget for the activity was under $300,000 or that the activities were part of a larger education investment that supported education in many fields other than STEM.

TABLE A1: CRITERIA AND DEFINITIONS

STEM: For the purposes of this inventory, STEM includes physical and natural sciences, technology, engineering, and mathematics disciplines, topics, or issues (including environmental science education or environmental stewardship). We recognize that various different and usually broader definitions are used for "STEM." This relatively narrow definition has been chosen to constrain the focus of the inventory to specific areas that have similar educational contexts, issues, and challenges, in order to maximize the inventory's usefulness in characterizing and improving the effectiveness of the Federal spending intended to address this particular set of educational contexts, issues, and challenges.

Investment (the unit of analysis in the detailed survey): A funded STEM education activity that has a dedicated budget of or above $300,000 (potentially part of a budget for a larger program, but excluding a one-time or irregular expenditure of overhead funds and salary and expenses costs), staff to manage the budget, and was funded in FY 2011.

STEM Education: Formal or informal, in-school or out-of-school, education that is primarily focused on physical and natural sciences, technology, engineering, and mathematics disciplines, topics, or issues (including environmental science education or environmental stewardship). All the investments included in this STEM education inventory have one of the following as a primary objective:

- **Learning:** Develop STEM skills, practices, or knowledge of students or the public.
- **Engagement:** Increase learners' interest in STEM, their perception of its value to their lives, or their ability to participate in STEM.
- **Pre- and In-Service Educator or Education Leader Performance:** Train or retain STEM educators (K–12 pre-service or in-service, postsecondary, and informal) and education leaders to improve their content knowledge and pedagogical skills.
- **Postsecondary STEM Degrees:** Increase the number of students who enroll in STEM majors, complete STEM credentials or degree programs, or are prepared to enter STEM careers or advanced education.
- **STEM Careers**: Prepare people to enter into the STEM workforce with training or certification (where STEM-discipline-specific knowledge and skill are the primary focus of the education investment).
- **STEM System Reform**: Improve STEM education through a focus on education system reform.
- **Institutional Capacity**: Support advancement and development of STEM personnel, programs, and infrastructure in educational institutions such as universities, informal education institutions, and state and local education agencies.
- **Education Research and Development:** Develop evidence-based STEM education models and practices.

Survey Structure

The inventory classifies Federal agency STEM education investments into two broad categories:

Agency-mission workforce education investments develop or train the STEM workforce of the agency or the workforce in fields directly related to the agency's mission (such as in aerospace engineering, national security science, and nuclear regulatory science). These typically include graduate scholarships, undergraduate internships, or institutional capacity-building in fields or degrees tightly aligned to an agency's mission.

Broader STEM education investments support formal and informal STEM education not focused on a specific discipline, STEM education research, and STEM education capacity building to improve interest in and understanding of STEM concepts and to enhance the broader national STEM workforce.

The FY 2011 survey contains five sections:

- Section 1: Background information (name, agency, and primary staff contact)
- Section 2: Descriptive information about broader STEM education investments (objectives, audiences, and numbers served)
- Section 3: Funding information about broader STEM education investments[138]
- Section 4: Evaluation information about broader STEM education investments
- Section 5: Information about agency-mission workforce education investments

Survey Data and Analysis

The FY 2011 inventory identified more than $2,891 million across more than 200 STEM education investments from 13 Federal agencies (the 12 CoSTEM member agencies and the NRC) (Figure A1). The investments include a range of programs and activities, such as student scholarships, grants to institutions, and contracts with STEM education organizations. Some investments also generate products and services for students and educators.

FIGURE A1: FEDERAL STEM EDUCATION INVESTMENTS >$300K, BY AGENCY (IN MILLIONS)[139]

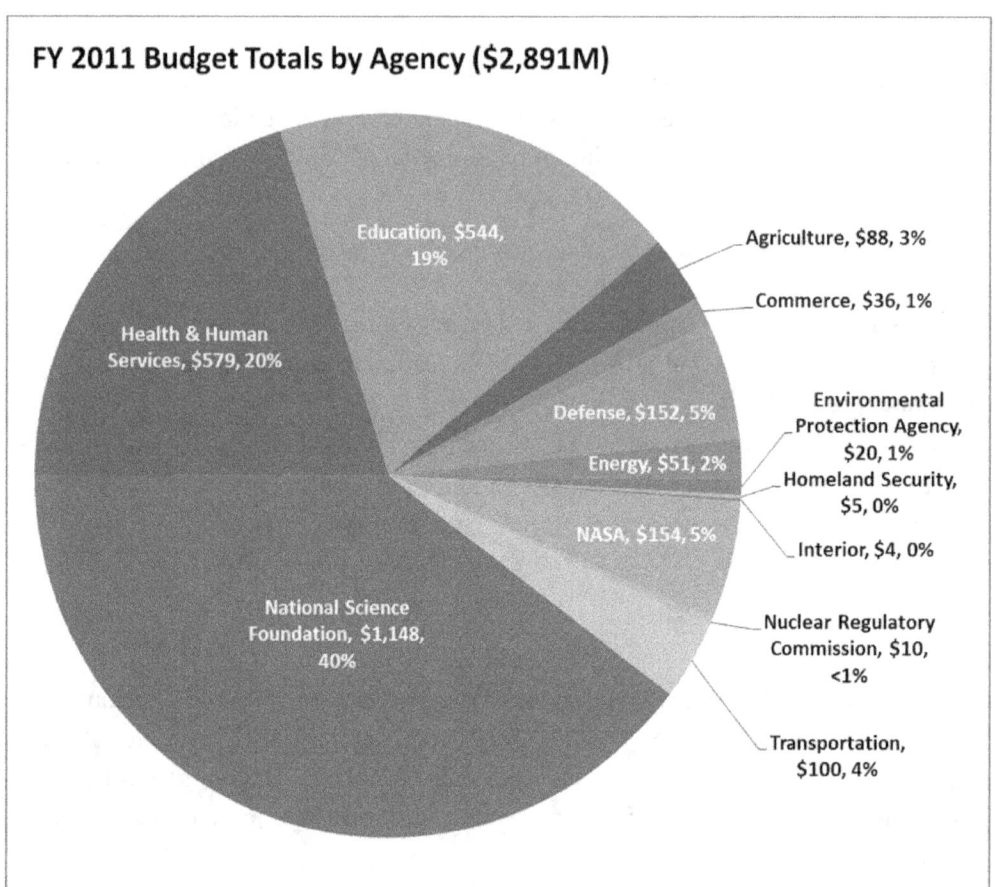

Investment Focus

Two-thirds of investments focus on broader STEM education

In FY 2011, Federal agencies spent $1,903 million on education investments focused on broader STEM education (66 percent of the nearly $2,891 million total) and $988 million (34 percent) on agency-mission workforce education.

The Federal STEM education portfolio reflects the agencies' missions and evolution of their roles in STEM education, which is influenced by the history and goals of education funding at each agency, emerging national needs, agency missions, Congressional and Presidential direction, and each agency's unique assets and capabilities. ED and NSF share the broadest missions in education and STEM education research respectively. ED and NSF together sponsor the majority of Federal investments in broader STEM education. Nearly all their investments focus on broader STEM issues and activities.

Most agencies focus the majority of their investments on agency-mission workforce issues

Science-mission agencies have broad responsibilities and make strategic investments in educating their workforces. Science-mission agencies devote between 55 and 100 percent of their STEM education investments to agency-mission workforce education.

Some agencies, including USDA, DOC, DOD, NOAA, and NASA, support both broader STEM education and agency-mission workforce education investments. Other agencies, such as DOE, HHS, and DOT, invest almost exclusively in mission-workforce development and training, whereas, DHS, DOI, and NRC allocate investments entirely to the education of a mission-related workforce.

K–12 education receives largest share of broader STEM investments

About $1.1 billion in broader STEM education investments were directed toward improving STEM education for K–12 students or teachers in FY 2011. About 60 percent of investments in broader STEM education focus on two or more types of education activities (e.g., K–12 education and undergraduate education; or informal education, K–12 education, and graduate education).

Primary Objectives

More than one-third of investments are aimed at helping students obtain postsecondary STEM degrees

Agencies identified a single primary objective (from a list of eight) for each investment (see Table A1). "Primary objective" was defined as the main desired outcome or basis for evaluating that investment under ideal circumstances. The Postsecondary STEM Degree primary objective accounted for 37 percent of all funding, followed by Education Research and Development (R&D; 16 percent) and STEM Careers (15 percent). Investments in STEM Careers and Postsecondary STEM Degrees are more likely to be related to agency-mission workforce education than to broader STEM education.

Funding for investments with the primary objective of Engagement was $164 million. In addition, engagement is the secondary objective of many investments. In fact, $1,444 million was spent on investments where Engagement was a secondary objective.[140]

About 70 percent of investments with the Postsecondary STEM Degrees primary objective were accounted for by five broader STEM education programs supported by NSF. They are the Graduate Research Fellowship Program; Scholarships in Science Technology, Engineering and Mathematics; Advanced Technological Education; Integrative Graduate Education and Research Traineeship Program; and Louis Stokes Alliances for Minority Participation. They included a range of investment types, from providing direct support for students seeking postsecondary STEM degrees, to providing support to 2-year and 4-year colleges, to supporting innovative models of graduate education and STEM-degree attainment among individuals from underrepresented groups. HHS investments accounted for nearly 40 percent of the funding for agency-mission workforce investments with the primary objective of Postsecondary STEM Degrees. HHS investments provided direct support to students or to 2-year and 4-year colleges to ensure sufficient graduates in biomedical fields.

$1,265 million in investments have an objective aimed at Educator Performance

Investments with the primary objective of Pre- and In-Service Educator and Education Leader Performance (Educator Performance) accounted for $315 million. Seventy-eight percent of those funds supported professional development (PD) for in-service educators, while the remainder supported both teacher PD and pre-service educators. No Federal investments are dedicated solely to funding efforts for pre-service educators. Educator Performance is listed as a secondary objective in investments that account for about $950 million.

Serving Groups that are Underrepresented and Underserved in STEM

Promoting equitable participation of people seeking STEM degrees and careers in STEM is a Federal priority and a common emphasis of investments in this survey. Information was collected on groups historically underrepresented in STEM fields, including Hispanics and Latinos, African Americans, American Indians, Alaska Natives, Native Hawaiians, and Pacific Islanders. In addition, information was collected on groups that may be disadvantaged or underserved in specific fields of STEM such as women, people with disabilities, economically disadvantaged people, and those living in rural or urban areas.

The FY 2011 survey used several questions to quantify investments that seek to enhance participation of people from underrepresented and underserved groups in STEM education. However, interpreting these data is difficult because agencies follow different regulations for allocating funds for programs directed toward specific groups. Further, statutory authority sometimes determines how, or whether, Federal agencies can legally limit specific investments to certain populations, such as through competitive priority or by setting aside specified funding amounts for specific populations.

Postsecondary Degrees and Institutional Capacity are top objectives for STEM equity

Eighty-nine percent of investment dollars were reported to emphasize or encourage support of underrepresented groups. Table A2 shows primary objectives of investments and type of focus. Of the $616 million in investments that focused primarily on underrepresented groups, 40 percent focused primarily on Postsecondary STEM Degrees ($248.5 million) and 33 percent focused on Institutional Capacity ($201.5 million). Postsecondary STEM Degrees was the most-funded of the eight primary objectives ($1,065 million), accounting for 37 percent of those investments.

Table A2: Distribution by Primary Objective of Investments Targeting Groups Underrepresented and Underserved in STEM, by Type of Focus on Underrepresented Groups (in millions)

Type of Focus on Underrepresented Groups	Education R&D	Engage-ment	Institutional Capacity	Learn-ing	STEM Degrees	Pre- and In-Service	STEM Careers	Total
Primary Focus	$27	$48	$201	$22	$249	$0	$68	$616
Specific Percent of Funding	$0	$8	$0	$0	$0	$0	$0	$8
Competitive Priority	$195	$7	$2	$16	$75	$1	$1	$296
Highly Encouraged	$228	$88	$9	$135	$610	$246	$328	$1,644
No Focus	$7	$13	$5	$71	$130	$69	$32	$327
Total	**$457**	**$164**	**$217**	**$245**	**$1,065**	**$315**	**$429**	**$2,891**

Institutional Capacity was a less common primary focus overall, yet 33 percent of investment dollars that focus primarily on underrepresented or underserved groups support building Institutional Capacity.

Minority Serving Institution Investments account for 10 percent of Federal STEM education funding

More than 20 Federal STEM education investments targeted historically unrepresented groups by supporting minority serving institutions (MSI). Overall, $297 million was directed towards MSIs. Twenty-eight percent can be awarded to any type of MSI. The remainder was targeted to specific types of MSIs, such as Historically Black Colleges and Universities (HBCU), Hispanic-Serving Institutions (HSI), Alaska Native-Serving Institutions, Native Hawaiian-Serving Institutions, and Tribal Colleges and Universities. The largest amount of funding ($110 million) was directed to HSIs, mostly from ED's Developing Hispanic-Serving Institutions STEM and Articulation Program ($100 million).

Audiences Served

Pre-K through undergraduate learners are the largest audience served

Federal STEM education investments served an array of audiences, and some served more than one. As illustrated in Table A3, the largest amount of funding ($2,354 million) is for investments that served learners in pre-kindergarten through graduate school. Investments that served educators and staff at formal education institutions are secondary in both number and amount of investments. Education researchers and informal-education educators and researchers received fewer funds; programs that served adults receive the least ($325 million).

TABLE A3: EDUCATION INVESTMENTS, BY TARGET GROUP (IN MILLIONS)[141]

Audience	FY 2011 Total
Learners Aged Pre-K to Grade 20	$2,354
K–12 Classroom Teachers	$1,059
K–12 Staff, Leaders, and Administrators	$416
Postsecondary Instructors	$837
Postsecondary Dean, Leaders, and Administrators	$371
Education Researchers	$386
Informal STEM Educators	$367
Informal STEM Education Leaders and Program Developers	$341
Adults[142]	$325

Interagency and Non-Federal Collaboration[143]

Thirty-two percent ($610 million) of the funding for broader STEM education supported collaborations with other Federal agencies or non-Federal groups. About $500 million supported interagency collaborations, including $123 million that supported collaboration across agencies and with non-federal organizations. Collaboration mechanisms include memoranda of understanding, interagency funding solicitations, and shared expertise.

Evaluation

Questions in both the FY 2010 and FY 2011 survey examined evaluations at the investment level. In FY 2011, 104 investments required grantees or investment components to conduct evaluations.

Of the more than 200 investments, 164 have been evaluated; 144 of those indicated the year the most recent evaluation was conducted. Since 2005, at least one evaluation has been completed for each of 138 investments (59 percent), with evaluations having been conducted on 44 percent in the past two years. A range of evaluation strategies was used at the investment level, and some investments were evaluated in more than one way.

Evaluation designs vary at the investment and component or grantee levels

For 109 investments their evaluation design was reported (Figure A2). Of those, the most common design was a pre-post intervention comparison with matched groups (46 investments, 29 percent). Randomized controlled trials were conducted for seven investments (4 percent). Fifty-five investments (34 percent) either did not report or did not specify an evaluation design. Forty-eight of the 109 investments used multiple evaluation designs, including four investments that used three different designs.

FIGURE A2: NUMBER OF EVALUATIONS, BY EVALUATION DESIGN (AT THE INVESTMENT LEVEL)[144]

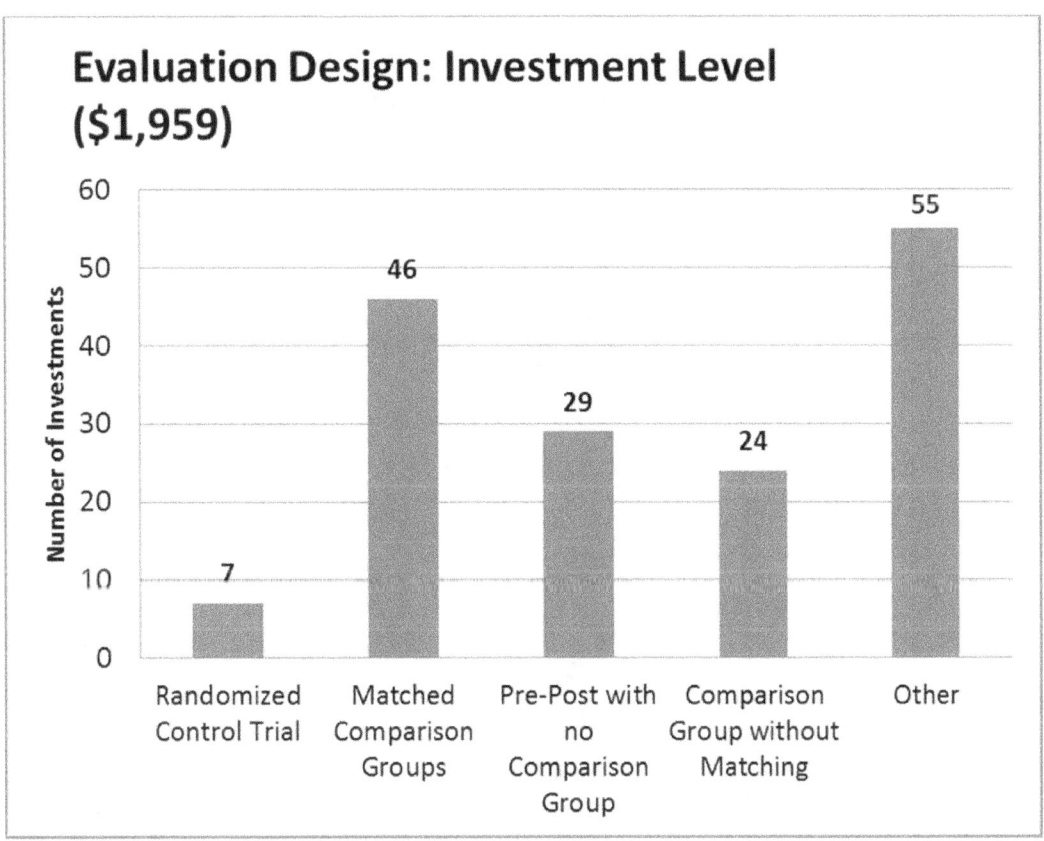

One-hundred-four investments required grantees or components to conduct evaluations. A variety of strategies were used at the investment component level, including those that identify how investments can be improved (formative) or test their impact (summative). Seventy-seven investments reported on the types of evaluation designs conducted at the level of the grantee or investment component (Figure A3). In contrast to evaluations at the investment level, the most common evaluation design at the grantee or component level was the pre-post assessment with no comparison group (45 investments). Few investments (11) reported evaluations using randomized designs at the grantee or investment or component level. Thirty-five of the 77 investments had components or grantees that used multiple evaluation designs, including 24 investments that reported using of 3 different evaluation designs.

FIGURE A3: NUMBER OF INVESTMENTS SINCE FY 2005, BY EVALUATION DESIGN (AT THE COMPONENT PART OR GRANTEE LEVEL)

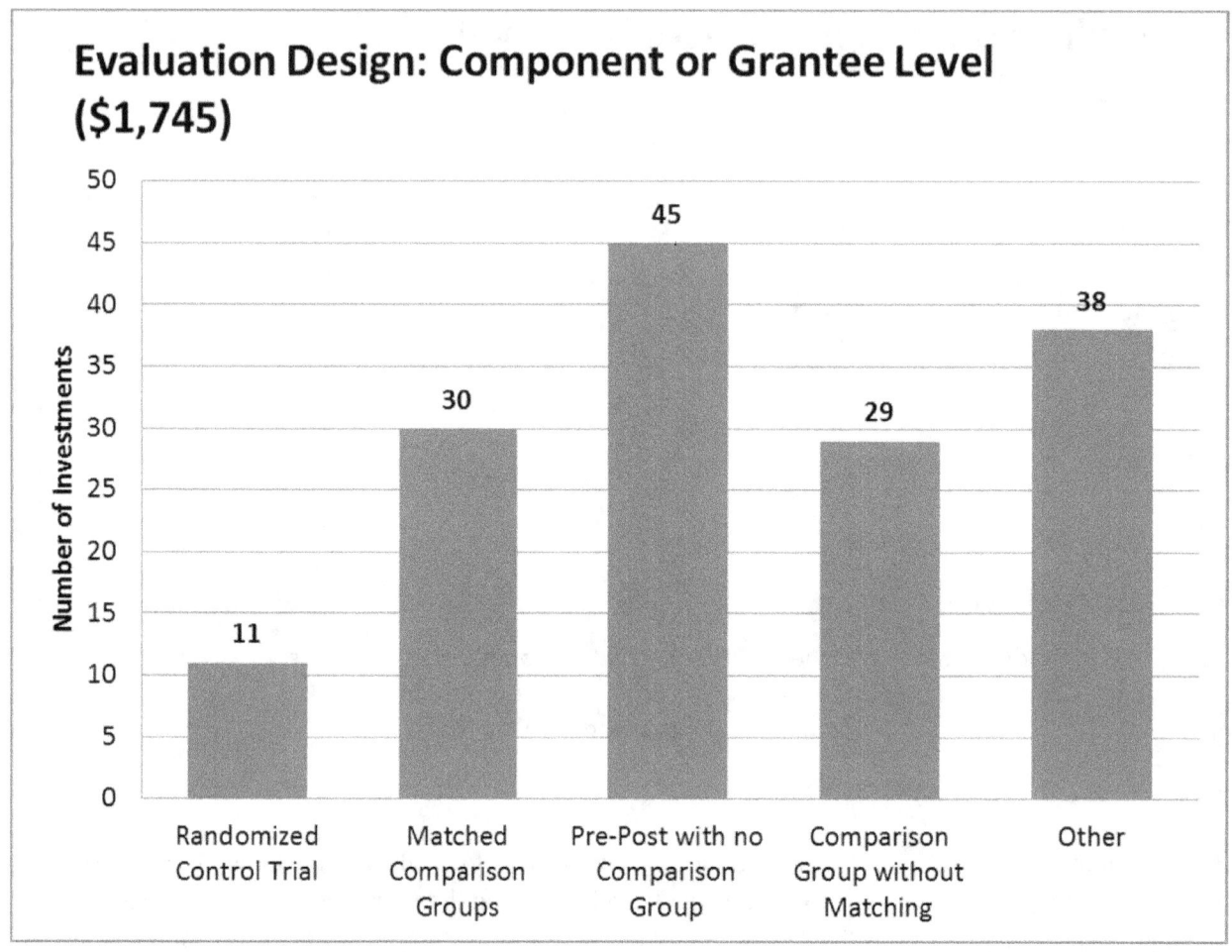

Impact of the Federal STEM Inventory Process and the FY 2010 and FY 2011 Portfolio Reports

Benefits from Collecting a STEM Education Inventory

The STEM education inventories of FY 2010 and FY 2011 have enabled agencies to respond better to inquiries about their STEM education investments. In addition, agencies have used inventory results to drive development or revisions of their own internal STEM education strategic planning.

The inventory has also led to changes in agency coordination and collaboration by catalyzing additional partnerships between and among agencies.

CoSTEM can help promote exchange of Federal agency staff expertise

The FY 2011 survey asked each program director to identify areas of expertise (evidence-based instructional practices and scaling up of effective efforts, for example) to share with colleagues across the Federal Government, as well as areas in which they would likely benefit from other expertise.

Identifying and connecting to subject matter experts (115 investments) was the most-frequently reported need cited by Federal staff who completed the survey, followed by design efforts for underrepresented or underserved groups (94 investments) and peer-review best practices (92 investments). Less frequently reported areas of expertise included evaluation of informal education (26 investments), evidence-based instructional practices for informal education (29 investments), and promotional campaigns for education products (30 investments).

Agency staff (representing 108 investments) most frequently expressed a desire to learn more about "Design Efforts for Underrepresented or Underserved Groups." That was followed by expertise on summative evaluations (99 investments) and the use of social media and social networking (95 investments). The least frequently cited areas of needed expertise included request for proposals development (26 investments), conducting proposal merit-reviews (28 investments), and running STEM competitions (32 investments).

Appendix Table A4: List of Investments in the FY 2011 Inventory

Agency	Investment	FY 08	FY 09	FY 10	FY 11	Type	Primary Objective	Under-represented Groups
Agriculture	1890 Institutions Capacity Building Grants Program: Extension	4.53	5.00	6.08	6.40	Broader STEM	Institutional Capacity	Yes
Agriculture	Agriculture in the Classroom	0.55	0.55	0.55	0.60	Broader STEM	Pre- & In-Service Educator Performance	No
Agriculture	Alaska Native-Serving and Native Hawaiian-Serving Institutions Education Competitive Grants Program	3.20	3.20	3.20	3.20	Agency Mission Workforce	Institutional Capacity	Yes
Agriculture	Resident Instruction Grants Program for Institutions of Higher Education in Insular Areas	0.75	0.80	0.90	0.86	Agency Mission Workforce	Institutional Capacity	Yes
Agriculture	Distance Education Grants for Institutions of Higher Education in Insular Areas (DEG)	-	-	0.75	0.70	Agency Mission Workforce	Institutional Capacity	Yes
Agriculture	1890 Institutions Capacity Building Grants Program: Teaching	4.53	5.00	6.08	6.40	Agency Mission Workforce	Institutional Capacity	Yes
Agriculture	1890 Facilities Grant Program	17.27	18.00	19.77	19.77	Agency Mission Workforce	Institutional Capacity	Yes
Agriculture	Hispanic-serving Institutions Education Grants Program	6.05	6.24	9.24	9.22	Agency Mission Workforce	Institutional Capacity	Yes
Agriculture	AITC Secondary Postsecondary Agriculture Education Challenge Grants (SPECA)	0.98	0.98	0.98	0.98	Agency Mission Workforce	Learning	No
Agriculture	Higher Education Challenge Grants (HEC)	5.39	5.65	5.65	5.64	Agency Mission Workforce	Learning	No

Agency	Investment	FY 08	FY 09	FY 10	FY 11	Type	Primary Objective	Under-represented Groups
Agriculture	New Era Rural Technology Competitive Grants Program (RTP)	-	0.75	0.88	0.87	Agency Mission Workforce	Postsecondary STEM Degrees	No
Agriculture	Higher Education Multicultural Scholars Program (MSP)	0.98	0.98	1.24	1.20	Agency Mission Workforce	Postsecondary STEM Degrees	Yes
Agriculture	Women and Minorities in Science, Technology, Engineering and Mathematics Fields Program (WAMS)	-	-	0.40	0.36	Agency Mission Workforce	STEM Careers	Yes
Agriculture	4-H Science, 4-H Youth Development Program	26.37	26.32	24.28	24.03	Broader STEM	Learning	No
Agriculture	National Institute of Food and Agriculture (NIFA) Fellowship Grants Program	6.00	6.00	6.50	3.00	Agency Mission Workforce	STEM Careers	No
Agriculture	Food and Agricultural Sciences National Needs Graduate and Postgraduate Fellowship Grant Program	3.68	3.86	3.86	3.90	Agency Mission Workforce	STEM Careers	Yes
Agriculture	AgDiscovery	0.49	0.49	0.49	0.50	Agency Mission Workforce	Engagement	No
Agriculture	1890 National Scholars Program	-	-	-	0.30	Agency Mission Workforce	Postsecondary STEM Degrees	Yes
Agriculture	**Total**	**80.75**	**83.82**	**90.85**	**87.94**			
Commerce	Satellite and Information Service	6.36	2.98	3.18	1.11	Broader STEM	Engagement	No
Commerce	NOAA Fisheries Education Program	2.24	2.24	2.31	3.46	Broader STEM	Engagement	No
Commerce	NOAA Bay Watershed Education and Training (B-WET)	9.55	9.70	9.70	2.85	Broader STEM	Engagement	Yes
Commerce	NOAA Teacher at Sea Program	0.19	0.60	0.60	0.60	Broader STEM	Pre- & In-Service Educator Performance	No

FEDERAL SCIENCE, TECHNOLOGY, ENGINEERING, AND MATHEMATICS (STEM) EDUCATION
STRATEGIC PLAN

Agency	Investment	FY 08	FY 09	FY 10	FY 11	Type	Primary Objective	Under-represented Groups
Commerce	NOAA Office of Ocean Exploration and Research (Education Only)	-	0.90	0.90	0.90	Broader STEM	Pre- & In-Service Educator Performance	No
Commerce	National Sea Grant College Program - Education Component	9.73	9.39	9.38	0.75	Broader STEM	Engagement	No
Commerce	Environmental Literacy Grants program	3.83	7.70	10.39	3.93	Broader STEM	Engagement	Yes
Commerce	National Estuarine Research Reserve System	2.82	3.50	4.00	1.99	Broader STEM	Engagement	No
Commerce	Educational Partnership Program with Minority Serving Institutions	13.92	14.98	14.31	14.27	Agency Mission Workforce	Postsecondary STEM Degrees	Yes
Commerce	NIST Summer Institute for Middle School Teachers	0.10	0.20	0.30	0.30	Broader STEM	Pre- & In-Service Educator Performance	No
Commerce	Coral Reef Conservation Program	0.83	0.83	0.84	0.50	Agency Mission Workforce	STEM Careers	Yes
Commerce	Summer Undergraduate Research Fellowship (SURF)	0.51	0.63	0.74	0.60	Agency Mission Workforce	Postsecondary STEM Degrees	No
Commerce	Ernest F. Hollings Undergraduate Scholarship Program	3.97	5.60	5.60	4.59	Agency Mission Workforce	Postsecondary STEM Degrees	Yes
Commerce	Dr. Nancy Foster Scholarship Program	0.47	0.62	0.60	0.51	Agency Mission Workforce	STEM Careers	Yes
Commerce	Total	54.51	59.87	62.85	36.37			
Defense	STARBASE	20.00	19.00	20.00	27.00	Broader STEM	Engagement	Yes
Defense	Navy - Historically Black Colleges and Universities/Minority Institutions Research and Education Partnership	1.50	1.50	1.50	1.50	Agency Mission Workforce	STEM Careers	Yes
Defense	University Laboratory Initiative (ULI)	2.60	2.30	2.35	2.30	Agency Mission Workforce	STEM Careers	No

Agency	Investment	FY 08	FY 09	FY 10	FY 11	Type	Primary Objective	Under-represented Groups
Defense	Air Force Consortium Research Fellows Program (CRFP)	-	-	-	0.81	Agency Mission Workforce	Postsecondary STEM Degrees	No
Defense	Army Educational Outreach Program (AEOP)	7.44	7.86	7.74	7.87	Broader STEM	Engagement	Yes
Defense	Iridescent Learning	-	-	0.61	0.86	Broader STEM	Engagement	Yes
Defense	Navy - Science and Engineering Apprenticeship Program (SEAP)	0.27	0.31	0.70	0.78	Broader STEM	Engagement	No
Defense	SeaPerch	0.45	0.75	0.90	1.45	Broader STEM	Engagement	Yes
Defense	The Naval Research Enterprise Intern Program (NREIP)	1.20	1.20	1.90	1.20	Agency Mission Workforce	Postsecondary STEM Degrees	No
Defense	Uniformed Services University of the Health Sciences (USUHS)	0.51	0.53	0.45	0.45	Agency Mission Workforce	STEM Careers	No
Defense	University NanoSatellite Program	1.50	1.60	1.60	1.60	Agency Mission Workforce	Learning	No
Defense	National Defense Education Program (NDEP) K-12 component	1.00	15.00	11.70	11.20	Broader STEM	Engagement	No
Defense	National Defense Education Program (NDEP) Science, Mathematics And Research for Transformation (SMART)	19.00	33.00	31.17	48.80	Agency Mission Workforce	Postsecondary STEM Degrees	No
Defense	National Science Center (NSC)	1.84	1.86	1.98	1.76	Broader STEM	Engagement	No
Defense	National Defense Science and Engineering Graduate (NDSEG) Fellowship Program	33.70	39.34	36.81	37.50	Agency Mission Workforce	Postsecondary STEM Degrees	No
Defense	Awards to Stimulate and Support Undergraduate Research Experiences (ASSURE)	4.50	4.50	4.50	4.50	Agency Mission Workforce	Postsecondary STEM Degrees	Yes
Defense	Stokes Educational Scholarship Program	2.07	1.93	1.94	2.16	Agency Mission Workforce	STEM Careers	Yes

Agency	Investment	FY 08	FY 09	FY 10	FY 11	Type	Primary Objective	Under-represented Groups
Defense	**Total**	**97.57**	**127.68**	**125.81**	**151.75**			
Education	Graduate Assistance in Areas of National Need (GAANN)	30.00	31.00	31.00	31.00	Broader STEM	Postsecondary STEM Degrees	No
Education	Teacher Loan Forgiveness	15.44	38.63	49.77	66.70	Broader STEM	Pre- & In-Service Educator Performance	No
Education	Minority Science and Engineering Improvement Program	8.58	8.58	9.50	9.48	Broader STEM	Postsecondary STEM Degrees	Yes
Education	Strengthening Predominantly Black Institutions	4.74	8.94	-	5.80	Broader STEM	Institutional Capacity	Yes
Education	Developing Hispanic Serving Institutions STEM and articulation programs	100.00	100.00	100.00	100.00	Broader STEM	Institutional Capacity	Yes
Education	Research, Development, and Dissemination	76.30	52.22	63.81	35.01	Broader STEM	Education R & D	No
Education	Research in Special Education	2.45	10.67	14.64	6.70	Broader STEM	Education R & D	No
Education	High School Longitudinal Study of 2009	6.45	5.14	6.15	6.50	Broader STEM	Education R & D	No
Education	Mathematics and Science Partnerships	179.00	179.00	180.50	175.10	Broader STEM	Pre- & In-Service Educator Performance	Yes
Education	Investing in Innovation	-	-	110.50	74.00	Broader STEM	Education R & D	Yes
Education	Upward Bound Math and Science Program	31.19	35.20	34.87	33.81	Broader STEM	Postsecondary STEM Degrees	Yes
Education	**Total**	**454.15**	**469.38**	**600.75**	**544.11**			
Energy	QuarkNet	0.75	0.75	0.75	0.75	Broader STEM	Pre- & In-Service Educator Performance	No
Energy	National Science Bowl	1.67	1.76	2.45	3.13	Broader STEM	Learning	No
Energy	Minority University Research Associates Program (MURA)	-	-	0.59	0.46	Agency Mission Workforce	STEM Careers	Yes

Agency	Investment	FY 08	FY 09	FY 10	FY 11	Type	Primary Objective	Under-represented Groups
Energy	Plasma/Fusion Science Educator Programs	0.73	0.77	0.77	0.77	Broader STEM	Pre- & In-Service Educator Performance	No
Energy	Advanced Vehicle Competitions	1.39	1.75	2.00	2.00	Agency Mission Workforce	STEM Careers	No
Energy	Graduate Automotive Technology Education	0.50	0.95	1.00	1.00	Agency Mission Workforce	Postsecondary STEM Degrees	No
Energy	Solar Decathlon	2.30	6.40	5.00	5.00	Agency Mission Workforce	Learning	No
Energy	Industrial Assessment Centers	3.40	3.30	6.10	4.10	Agency Mission Workforce	STEM Careers	No
Energy	Visiting Faculty Program (formerly FaST)	0.25	1.54	1.02	0.61	Agency Mission Workforce	STEM Careers	Yes
Energy	Wind for Schools	0.37	0.46	0.63	0.60	Agency Mission Workforce	Postsecondary STEM Degrees	No
Energy	Minority Educational Institution Student Partnership Program	0.55	0.66	1.50	1.50	Agency Mission Workforce	Engagement	Yes
Energy	Special Recruitment Programs/Mickey Leland Fellowship	0.70	0.70	0.70	0.70	Agency Mission Workforce	Engagement	Yes
Energy	HBCU Mathematics, Science & Technology, Engineering and Research Workforce Development Program	-	-	8.97	8.27	Agency Mission Workforce	STEM Careers	Yes
Energy	National Undergraduate Fellowship Program in Plasma Physics and Fusion Energy Sciences	0.37	0.37	0.37	0.37	Agency Mission Workforce	STEM Careers	No

Agency	Investment	FY 08	FY 09	FY 10	FY 11	Type	Primary Objective	Under-represented Groups
Energy	Global Change Education Program	1.47	1.42	1.42	0.88	Agency Mission Workforce	Postsecondary STEM Degrees	No
Energy	American Chemical Society Summer School in Nuclear and Radiochemistry	0.52	0.53	0.55	0.56	Agency Mission Workforce	Learning	No
Energy	Computational Science Graduate Fellowship	6.80	6.80	7.80	6.00	Agency Mission Workforce	Postsecondary STEM Degrees	No
Energy	Office of Science Graduate Fellowship (SCGF) program	-	-	5.00	5.00	Agency Mission Workforce	STEM Careers	No
Energy	Science Undergraduate Laboratory Internships	2.58	2.50	3.80	6.34	Agency Mission Workforce	STEM Careers	No
Energy	Community College Internship	0.32	0.29	0.69	0.66	Agency Mission Workforce	STEM Careers	No
Energy	Hydro Research Fellowships	-	0.90	0.90	2.35	Agency Mission Workforce	STEM Careers	No
Energy	Total	24.66	31.86	52.00	51.05			
Environmental Protection Agency	Environmental Education Grants	3.40	3.40	2.20	3.70	Broader STEM	Learning	No
Environmental Protection Agency	National Environmental Education Training Program	2.00	2.00	2.20	2.40	Broader STEM	Pre- & In-Service Educator Performance	Yes
Environmental Protection Agency	P3-People, Prosperity & the Planet-Award: A National Student Design Competition for Sustainability	1.30	1.20	2.00	1.60	Broader STEM	Engagement	No

Agency	Investment	FY 08	FY 09	FY 10	FY 11	Type	Primary Objective	Under-represented Groups
Environmental Protection Agency	Cooperative Training Partnership in Environmental Sciences Research	2.00	1.50	1.50	0.50	Agency Mission Workforce	STEM Careers	No
Environmental Protection Agency	Greater Research Opportunities (GRO) Fellowships for Undergraduate Environmental Study	0.60	1.30	1.50	2.00	Agency Mission Workforce	Postsecondary STEM Degrees	No
Environmental Protection Agency	Science to Achieve Results Graduate Fellowship Program	8.22	4.24	6.39	9.50	Agency Mission Workforce	Postsecondary STEM Degrees	No
Environmental Protection Agency	University of Cincinnati/EPA Research Training Grant	0.60	0.60	0.60	0.60	Agency Mission Workforce	STEM Careers	No
	Total	**18.12**	**14.24**	**16.39**	**20.30**			
Health & Human Services	Science Education Drug Abuse Partnership Award	2.16	2.46	2.30	1.90	Broader STEM	Learning	Yes
Health & Human Services	NIAID Science Education Awards	0.35	0.71	1.06	0.61	Broader STEM	Learning	No
Health & Human Services	NIH Science Education Partnership Award (SEPA)	16.18	22.21	18.32	18.90	Broader STEM	Learning	Yes
Health & Human Services	Curriculum Supplement Series	0.76	0.36	0.34	0.35	Broader STEM	Learning	No
Health & Human Services	Mathematics and Science Cognition and Learning (MSCL) Program	4.70	10.10	10.40	9.20	Broader STEM	Education R & D	No
Health & Human Services	Office of Science Education K-12 Program	2.10	2.11	2.27	2.27	Broader STEM	Engagement	Yes
Health & Human Services	NLM Institutional Grants for Research Training in Biomedical Informatics	10.33	14.66	10.14	12.20	Postsecondary STEM Degrees	Postsecondary STEM Degrees	Yes

Agency	Investment	FY 08	FY 09	FY 10	FY 11	Type	Primary Objective	Under-represented Groups
Health & Human Services	Research Supplements to Promote Diversity in Health-Related Research	70.56	83.43	68.98	66.20	Agency Mission Workforce	Postsecondary STEM Degrees	Yes
Health & Human Services	Ruth L. Kirschstein National Research Service Award Institutional Research Training Grants (T32, T35)	259.08	266.47	230.84	219.40	Agency Mission Workforce	STEM Careers	Yes
Health & Human Services	Ruth L. Kirschstein NRSA for Individual Predoctoral Fellows, including Underrepresented Racial/Ethnic Groups, Students from Disadvantaged Backgrounds, and Predoctoral Students with Disabilities	47.57	55.55	56.88	54.10	Agency Mission Workforce	STEM Careers	Yes
Health & Human Services	MARC U-STAR NRSA Program	16.76	21.25	21.25	20.90	Agency Mission Workforce	Postsecondary STEM Degrees	Yes
Health & Human Services	Short Courses on Mathematical, Statistical, and Computational Tools for Studying Biological Systems	0.32	0.33	0.70	0.90	Agency Mission Workforce	Learning	No
Health & Human Services	Short Courses in Integrative and Organ Systems Pharmacology	0.68	0.75	0.67	0.67	Agency Mission Workforce	Learning	No
Health & Human Services	Initiative for Maximizing Student Development	16.44	22.34	21.41	22.30	Agency Mission Workforce	Postsecondary STEM Degrees	Yes
Health & Human Services	RISE (Research Initiative for Scientific Enhancement)	18.57	25.69	24.44	26.20	Agency Mission Workforce	Postsecondary STEM Degrees	Yes
Health & Human Services	Bridges to the Baccalaureate Program	4.00	7.26	6.46	6.71	Agency Mission Workforce	Postsecondary STEM Degrees	Yes
Health & Human Services	Bridges to the Doctorate	1.29	2.18	2.98	3.17	Agency Mission Workforce	Postsecondary STEM Degrees	Yes

Agency	Investment	FY 08	FY 09	FY 10	FY 11	Type	Primary Objective	Under-represented Groups
Health & Human Services	Postbaccalaureate Research Education Program (PREP)	3.03	6.73	5.78	7.70	Agency Mission Workforce	Postsecondary STEM Degrees	Yes
Health & Human Services	Cancer Education Grants Program	5.75	7.31	6.76	5.70	Agency Mission Workforce	Learning	No
Health & Human Services	National Cancer Institute Cancer Education and Career Development Program	15.21	18.99	20.44	18.40	Agency Mission Workforce	Learning	No
Health & Human Services	Summer Institute for Training in Biostatistics	-	1.45	1.45	1.50	Agency Mission Workforce	Postsecondary STEM Degrees	No
Health & Human Services	Short-Term Research Education Program to Increase Diversity in Health-Related Research	2.05	3.17	4.19	4.00	Agency Mission Workforce	Postsecondary STEM Degrees	Yes
Health & Human Services	Short Term Educational Experiences for Research (STEER) in the Environmental health Sciences for Undergraduates and High School Students	0.57	0.76	0.57	0.50	Agency Mission Workforce	Engagement	No
Health & Human Services	NINDS Diversity Research Education Grants in Neuroscience	0.25	0.71	0.82	1.00	Agency Mission Workforce	Postsecondary STEM Degrees	Yes
Health & Human Services	Short Courses in Population Research (Education Programs for Population Research R25)	0.70	0.92	0.75	0.80	Agency Mission Workforce	Learning	No
Health & Human Services	Graduate Program Partnerships	17.60	16.60	16.70	16.00	Agency Mission Workforce	STEM Careers	No
Health & Human Services	Post-baccalaureate Intramural Research Training Award Program	21.54	21.48	24.81	23.50	Agency Mission Workforce	Postsecondary STEM Degrees	No
Health & Human Services	Student Intramural Research Training Award Program	5.78	5.12	5.87	5.40	Agency Mission Workforce	Learning	No

Agency	Investment	FY 08	FY 09	FY 10	FY 11	Type	Primary Objective	Under-represented Groups
Health & Human Services	Technical Intramural Research Training Award	2.14	2.14	2.24	2.24	Agency Mission Workforce	STEM Careers	No
Health & Human Services	Undergraduate Scholarship Program for Individuals from Disadvantaged Backgrounds	2.20	2.30	2.40	2.50	Agency Mission Workforce	Postsecondary STEM Degrees	Yes
Health & Human Services	Clinical Research Training Program	1.00	1.00	1.10	0.85	Agency Mission Workforce	Learning	No
Health & Human Services	Health Careers Opportunity Program	-	-	-	22.00	Broader STEM	Postsecondary STEM Degrees	Yes
Health & Human Services	Public Health Traineeship	-	-	-	1.40	Agency Mission Workforce	STEM Careers	No
Total		**551.66**	**626.52**	**573.32**	**579.45**			
Homeland Security	National Nuclear Forensics Expertise Development Program	-	-	-	2.50	Agency Mission Workforce	STEM Careers	Yes
Homeland Security	Homeland Security STEM Career Development Grant Program	3.50	2.50	2.50	1.00	Agency Mission Workforce	STEM Careers	No
Homeland Security	Scientific Leadership Awards Program	3.80	3.80	3.40	1.00	Agency Mission Workforce	STEM Careers	Yes
Homeland Security Total		**7.30**	**6.30**	**5.90**	**4.50**			
Interior	Educational Component of the National Geologic Mapping Program (EDMAP)	0.49	0.52	0.57	0.57	Agency Mission Workforce	Postsecondary STEM Degrees	No
Interior	Conservation and Land Management Internship Program	-	-	-	1.74	Agency Mission Workforce	STEM Careers	Yes

Agency	Investment	FY 08	FY 09	FY 10	FY 11	Type	Primary Objective	Under-represented Groups
Interior	George Melendez Wright Climate Change Youth Initiative	-	-	-	0.91	Agency Mission Workforce	STEM Careers	Yes
Interior	Geoscientists–n-the-Parks Program	-	-	-	0.33	Agency Mission Workforce	Learning	Yes
Interior	**Total**	**0.49**	**0.52**	**0.57**	**3.55**			
NASA	LDCM	0.14	0.54	0.30	0.68	Broader STEM	Pre- & In-Service Educator Performance	Yes
NASA	HST	1.42	1.35	1.25	1.29	Broader STEM	Learning	Yes
NASA	Cassini	1.55	1.70	1.65	1.26	Broader STEM	Learning	Yes
NASA	Aqua	-	0.48	0.43	0.44	Broader STEM	Engagement	No
NASA	GCCE - Global Climate Change Education	7.00	10.00	10.00	3.50	Broader STEM	Learning	Yes
NASA	Astrophysics Forum	-	0.99	1.00	1.02	Broader STEM	Institutional Capacity	Yes
NASA	Heliophysics	-	0.82	0.73	0.79	Broader STEM	Institutional Capacity	Yes
NASA	Planetary Science E/PO Forum	-	0.92	0.89	0.87	Broader STEM	Institutional Capacity	Yes
NASA	Earth Science E/PO Forum	-	0.76	0.87	0.89	Broader STEM	Institutional Capacity	Yes
NASA	Chandra	1.92	1.85	1.82	1.77	Broader STEM	Engagement	No
NASA	LTP - Learning Technologies Project	1.28	0.84	0.71	0.54	Broader STEM	Education R & D	No
NASA	EPOESS	0.79	4.58	6.91	7.00	Broader STEM	Learning	Yes
NASA	Juno	0.58	1.21	1.31	1.27	Broader STEM	Learning	Yes
NASA	DAWN	0.30	0.27	0.36	0.53	Broader STEM	Learning	Yes
NASA	GRAIL	0.20	0.31	0.41	0.53	Broader STEM	Engagement	Yes
NASA	NAS - NASA Aerospace Scholars			0.30	0.30	Broader STEM	Engagement	No
NASA	NES - NASA Explorer Schools	8.09	4.31	4.99	3.10	Broader STEM	Engagement	No
NASA	NASA's Beginning Engineering, Science and Technology (BEST) students	-	-	-	0.43	Broader STEM	Education R & D	Yes

Agency	Investment	FY 08	FY 09	FY 10	FY 11	Type	Primary Objective	Under-represented Groups
NASA	SOFIA (Stratospheric Observatory for Infrared Astronomy) Education and Public Outreach	0.24	0.36	0.60	0.61	Broader STEM	Pre- & In-Service Educator Performance	Yes
NASA	NSTI-MI - NASA Science and Technology Institute for Minority Institutions	2.00	1.96	2.46	2.11	Agency Mission Workforce	Institutional Capacity	Yes
NASA	Space Grant - National Space Grant College and Fellowship Program	39.71	38.30	44.50	45.50	Agency Mission Workforce	Postsecondary STEM Degrees	Yes
NASA	eEducation Small Projects/Central Operation of Resources for Educators (CORE)	0.60	0.49	0.40	0.49	Broader STEM	Pre- & In-Service Educator Performance	Yes
NASA	NETS - NASA Education Technologies Services	1.40	1.30	1.00	0.47	Broader STEM	Learning	Yes
NASA	Curriculum Improvement Partnership Award for the Integration of Research into the Undergraduate Curriculum (CIPAIR)	2.75	2.71	3.11	0.60	Agency Mission Workforce	Postsecondary STEM Degrees	Yes
NASA	University student launch initiative	-	-	-	0.32	Agency Mission Workforce	Engagement	Yes
NASA	Aeronautics Scholarship	1.80	1.80	1.80	1.60	Agency Mission Workforce	Postsecondary STEM Degrees	No
NASA	Innovation in Aeronautics Instruction Competition	1.10	1.10	1.10	0.46	Agency Mission Workforce	Institutional Capacity	No
NASA	JPFP - Jenkins Pre-Doctoral Fellowship Program	2.56	2.53	2.63	3.40	Agency Mission Workforce	Postsecondary STEM Degrees	Yes
NASA	USRP - Undergraduate Student Research Project	4.00	3.48	2.97	3.30	Agency Mission Workforce	Postsecondary STEM Degrees	No
NASA	EFP - Education Flight Projects	1.20	3.11	2.99	2.70	Broader STEM	Engagement	No
NASA	SEMAA - Science Engineering Mathematics and Aerospace Academy	2.51	1.91	3.09	2.13	Broader STEM	Learning	Yes

Agency	Investment	FY 08	FY 09	FY 10	FY 11	Type	Primary Objective	Under-represented Groups
NASA	SEED - Systems Engineering Educational Discovery	0.29	0.37	0.41	0.30	Agency Mission Workforce	STEM Careers	No
NASA	Reduced Gravity Student Flight Opportunity Project	-	0.36	0.36	0.30	Agency Mission Workforce	Learning	No
NASA	SOI - Summer of Innovation	-	-	10.00	10.10	Broader STEM	Engagement	Yes
NASA	HEOMD Space Grant Project	1.36	1.55	1.03	1.07	Agency Mission Workforce	Postsecondary STEM Degrees	Yes
NASA	NSBRI Higher Education Activities - National Space Biomedical Research Institute	0.72	0.74	0.75	0.65	Agency Mission Workforce	STEM Careers	Yes
NASA	MESSENGER	0.43	0.36	0.30	0.34	Broader STEM	Learning	No
NASA	URC - University Research Centers	13.93	14.57	14.06	10.40	Agency Mission Workforce	Postsecondary STEM Degrees	Yes
NASA	Aura	-	0.38	0.37	0.48	Broader STEM	Learning	No
NASA	GSRP - Graduate Student Researchers Program	5.20	4.30	4.40	3.30	Agency Mission Workforce	STEM Careers	No
NASA	Innovation in Higher Education STEM Education	-	-	0.96	4.70	Agency Mission Workforce	Postsecondary STEM Degrees	No
NASA	AESP - Aerospace Education Services Project	4.90	5.50	2.50	3.90	Broader STEM	Pre- & In-Service Educator Performance	Yes
NASA	Mars E/PO Informal Ed	0.99	0.77	0.81	1.00	Broader STEM	Engagement	No
NASA	Mars E/PO Formal Ed	1.20	1.00	1.30	1.10	Broader STEM	Learning	No
NASA	GLOBE (Learning and Observations to Benefit the Environment) Program		4.40	3.00	5.00	Broader STEM	Engagement	No
NASA	TCUP - NASA Tribal College and University Project	1.62	1.68	1.59	1.50	Agency Mission Workforce	Postsecondary STEM Degrees	Yes

Agency	Investment	FY 08	FY 09	FY 10	FY 11	Type	Primary Objective	Under-represented Groups
NASA	LEARN - Learning Environment and Research Network	2.40	3.00	3.00	2.70	Broader STEM	Engagement	No
NASA	SIMulation-Based Engineering and Science Teacher Professional Development	-	-	0.39	0.40	Broader STEM	Pre- & In-Service Educator Performance	No
NASA	INSPIRE - Interdisciplinary National Science Program Incorporating Research and Education Experience	2.85	3.42	2.52	3.23	Broader STEM	Engagement	Yes
NASA	LERCIP - Lewis Educational Research Collaborative Internship Project (College)	0.76	0.97	0.90	0.60	Agency Mission Workforce	Postsecondary STEM Degrees	No
NASA	MSP - MUREP Small Projects	1.50	1.80	1.70	1.80	Agency Mission Workforce	Institutional Capacity	Yes
NASA	LARSS - NASA Langley Aerospace Research Student Scholars Program	1.00	1.10	1.30	1.30	Agency Mission Workforce	Postsecondary STEM Degrees	No
NASA	MUST - Motivating Undergraduates in Science and Technology	1.90	1.90	2.40	1.80	Agency Mission Workforce	Postsecondary STEM Degrees	Yes
NASA	CEP - Career Exploration Project	1.16	1.15	1.30	1.20	Agency Mission Workforce	Engagement	Yes
NASA	Space technology research fellowships	-	-	-	7.00	Agency Mission Workforce	Learning	No
NASA	**Total**	**125.33**	**139.27**	**155.91**	**154.07**			
National Science Foundation	Opportunities for Enhancing Diversity in the Geosciences	4.57	11.79	4.18	3.60	Broader STEM	Engagement	Yes
National Science Foundation	Geoscience Education	1.63	2.74	2.02	1.50	Broader STEM	Learning	Yes
National Science Foundation	Climate Change Education (CCE)	-	9.95	10.24	10.00	Broader STEM	Learning	Yes

Agency	Investment	FY 08	FY 09	FY 10	FY 11	Type	Primary Objective	Under-represented Groups
National Science Foundation	CITEAM	5.90	0.00	4.85	5.00	Broader STEM	Learning	Yes
National Science Foundation	CISE Pathways to Revitalized Undergraduate Computing Education (CPATH)	5.00	5.00	4.37	1.50	Broader STEM	Institutional Capacity	Yes
National Science Foundation	Geoscience Teacher Training (GEO-Teach)	3.00	3.00	2.98	2.00	Broader STEM	Pre- & In-Service Educator Performance	Yes
National Science Foundation	Broadening Participation in Computing (BPC)	14.00	14.00	14.00	14.00	Broader STEM	Postsecondary STEM Degrees	Yes
National Science Foundation	Global Learning and Observations to Benefit the Environment (GLOBE)	1.10	1.12	1.10	1.10	Broader STEM	Engagement	No
National Science Foundation	Undergraduate Research and Mentoring in the Biological Sciences (URM)	5.09	4.68	9.00	3.00	Broader STEM	Engagement	Yes
National Science Foundation	Transforming Undergraduate Biology Education (TUBE)	-	0.90	5.06	15.00	Broader STEM	Engagement	Yes
National Science Foundation	Computing Education for the 21st Century (CE21)	-	-	-	5.50	Broader STEM	Learning	Yes
National Science Foundation	Graduate STEM Fellows in K-12 Education (GK-12)	54.60	58.84	55.97	52.95	Broader STEM	Learning	Yes
National Science Foundation	Math and Science Partnership (MSP)	47.87	85.99	57.93	57.08	Broader STEM	Education R & D	Yes
National Science Foundation	Robert Noyce Scholarship (Noyce) Program	55.05	115.00	54.93	54.89	Broader STEM	Pre- & In-Service Educator Performance	Yes
National Science Foundation	Science, Technology, Engineering, and Mathematics Talent Expansion Program (STEP)	29.48	29.09	31.64	33.44	Broader STEM	Postsecondary STEM Degrees	Yes
National Science Foundation	Enhancing the Mathematical Sciences Workforce in the 21st Century (EMSW21)	19.46	26.95	15.07	16.47	Broader STEM	Postsecondary STEM Degrees	Yes
National Science Foundation	Interdisciplinary Training for Undergraduates in Biological and Mathematical Sciences (UBM)	2.32	2.71	2.70	2.10	Broader STEM	Postsecondary STEM Degrees	Yes

FEDERAL SCIENCE, TECHNOLOGY, ENGINEERING, AND MATHEMATICS (STEM) EDUCATION
STRATEGIC PLAN

Agency	Investment	FY 08	FY 09	FY 10	FY 11	Type	Primary Objective	Under-represented Groups
National Science Foundation	International Research Experiences for Students (IRES)	2.71	4.21	3.43	3.15	Broader STEM	Postsecondary STEM Degrees	Yes
National Science Foundation	Scholarships in Science, Technology, Engineering, and Mathematics (S-STEM)	92.40	61.22	75.96	75.00	Broader STEM	Postsecondary STEM Degrees	Yes
National Science Foundation	Centers for Ocean Sciences Education Excellence	5.24	7.19	5.70	5.24	Broader STEM	Institutional Capacity	No
National Science Foundation	Advanced Technological Education (ATE)	51.46	51.85	64.51	64.00	Broader STEM	Postsecondary STEM Degrees	Yes
National Science Foundation	Research and Evaluation on Education in Science and Engineering (REESE)	41.66	42.60	45.67	45.59	Broader STEM	Education R & D	No
National Science Foundation	Research in Disabilities Education (RDE)	5.93	6.88	6.92	6.48	Broader STEM	Postsecondary STEM Degrees	Yes
National Science Foundation	Informal Science Education (ISE)	64.45	65.72	65.85	64.22	Broader STEM	Education R & D	Yes
National Science Foundation	Louis Stokes Alliances for Minority Participation (LSAMP)	40.47	42.50	44.55	45.63	Broader STEM	Postsecondary STEM Degrees	Yes
National Science Foundation	Transforming Undergrad Education in STEM (TUES)	37.28	40.86	41.60	40.92	Broader STEM	Learning	Yes
National Science Foundation	Excellence Awards in Science and Engineering (EASE)	5.57	5.15	5.18	5.18	Broader STEM	Engagement	Yes
National Science Foundation	Research Experiences for Undergraduates (REU)	62.67	100.47	80.99	63.55	Broader STEM	STEM Careers	Yes
National Science Foundation	Nanotechnology Undergraduate Education in Engineering	1.08	2.00	1.83	1.50	Broader STEM	Learning	Yes
National Science Foundation	Research Experiences for Teachers (RET) in Engineering and Computer Science	3.97	5.79	5.41	4.00	Broader STEM	Pre- & In-Service Educator Performance	Yes
National Science Foundation	Engineering Education (EE)	11.50	22.89	13.74	10.85	Broader STEM	Education R & D	Yes
National Science Foundation	Alliances for Graduate Education and the Professoriate (AGEP)	15.85	17.18	16.73	16.70	Broader STEM	Education R & D	Yes
National Science Foundation	Tribal Colleges and Universities Program (TCUP)	12.80	13.39	13.35	13.31	Broader STEM	Institutional Capacity	Yes

Agency	Investment	FY 08	FY 09	FY 10	FY 11	Type	Primary Objective	Under-represented Groups
National Science Foundation	Historically Black Colleges and Universities Undergraduate Program (HBCU-UP)	29.74	31.13	32.06	31.94	Broader STEM	Institutional Capacity	Yes
National Science Foundation	Graduate Research Fellowship Program (GRFP)	96.02	162.44	136.13	137.68	Broader STEM	Postsecondary STEM Degrees	Yes
National Science Foundation	Integrative Graduate Education and Research Traineeship (IGERT) Program	64.76	77.99	69.70	60.76	Broader STEM	Postsecondary STEM Degrees	Yes
National Science Foundation	East Asia & Pacific Summer Institutes for U.S. Graduate Students (EAPSI)	1.75	1.52	1.74	2.40	Broader STEM	Engagement	Yes
National Science Foundation	Federal Cyber Service: Scholarship for Service (SFS)	11.37	14.88	14.87	14.96	Agency Mission Workforce	STEM Careers	Yes
National Science Foundation	Discovery Research K-12 (DR-K12)	99.25	108.41	118.38	119.90	Broader STEM	Education R & D	Yes
National Science Foundation	Innovative Technology Experiences for Students and Teachers (ITEST)	28.72	27.86	20.85	25.00	Broader STEM	Engagement	Yes
National Science Foundation	Research on Gender in Science and Engineering (GSE)	10.13	11.40	11.57	10.47	Broader STEM	Education R & D	Yes
National Science Foundation Total		1049.85	1297.29	1172.76	1147.57			
Nuclear Regulatory Commission	Nuclear Education Curriculum Development Program	4.72	4.72	4.70	4.70	Agency Mission Workforce	Institutional Capacity	No
Nuclear Regulatory Commission	Minority Serving Institutions Program (MSIP)	1.00	1.42	2.84	0.70	Agency Mission Workforce	Postsecondary STEM Degrees	Yes
Nuclear Regulatory Commission	Integrated University Program - Scholarship and Fellowship Program	-	8.40	15.00	5.00	Agency Mission Workforce	Postsecondary STEM Degrees	No
Total		5.72	14.54	22.54	10.40			

Agency	Investment	FY 08	FY 09	FY 10	FY 11	Type	Primary Objective	Under-represented Groups
Transportation	Air Transportation Centers of Excellence (COE)	13.20	14.10	16.40	12.50	Agency Mission Workforce	Postsecondary STEM Degrees	No
Transportation	University Transportation Centers (UTC) Program	74.44	83.45	83.67	80.00	Agency Mission Workforce	Postsecondary STEM Degrees	No
Transportation	Dwight David Eisenhower Transportation Fellowship Program	1.96	1.99	2.01	2.00	Agency Mission Workforce	Postsecondary STEM Degrees	No
Transportation	Garrett A. Morgan Technology and Transportation Education Program (GAM)	1.11	1.13	1.14	1.10	Agency Mission Workforce	Learning	Yes
Transportation	Summer Transportation Institute Program for Diverse Groups (STIPDG)	0.64	0.65	0.65	3.90	Agency Mission Workforce	Engagement	Yes
Transportation	National Summer Transportation Institute Program (STI)	-	-	-	0.80	Agency Mission Workforce	Engagement	Yes
Transportation	**Total**	91.36	101.32	103.87	100.30			
	Grand Total	2561.48	2975.60	2983.56	2891.34			

Figure A4. FY 2014 President's Budget Request, Federal STEM Education Investments by Agency

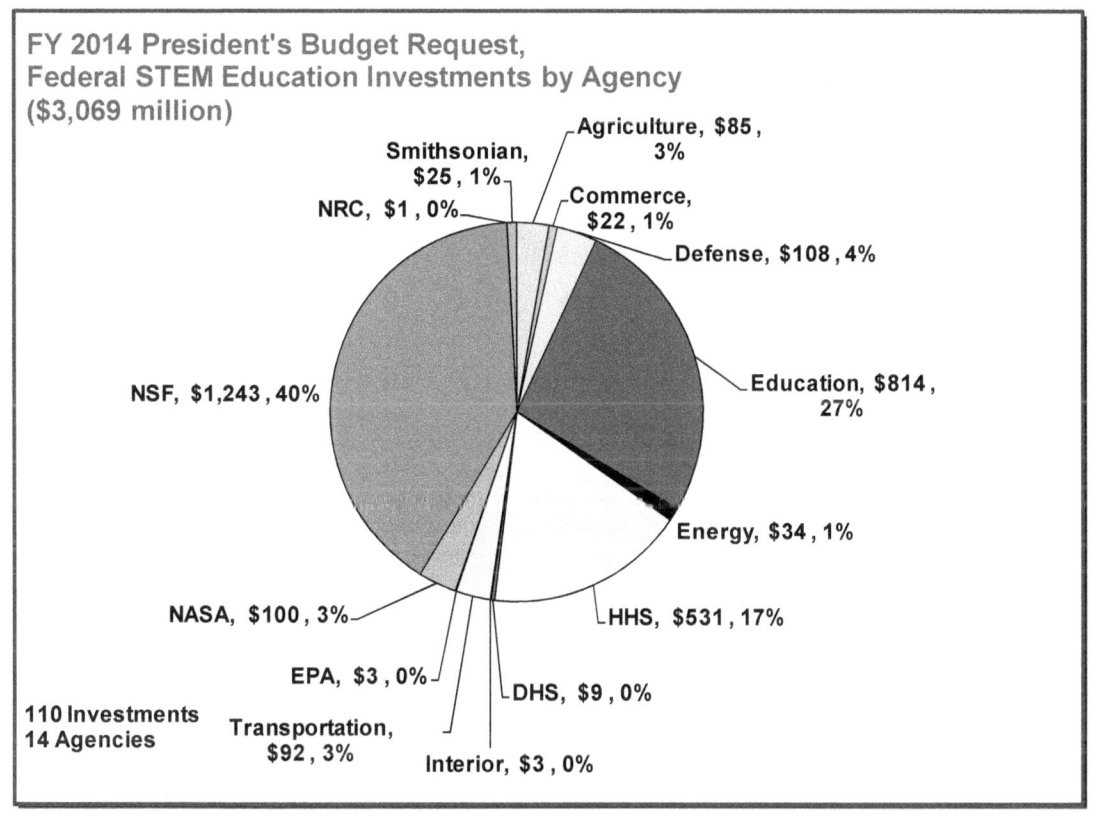

TABLE A5: FEDERAL STEM EDUCATION FUNDING BY AGENCY IN MILLIONS FY 2012 AND2014

	2012 Enacted	2014 Requested
Agriculture	$ 88	$ 85
Commerce	$ 40	$ 22
Defense	$ 153	$ 108
DHS	$ 11	$ 9
Education	$ 529	$ 814
Energy	$ 47	$ 34
EPA	$ 26	$ 3
HHS	$ 576	$ 531
Interior	$ 3	$ 3
NASA	$ 149	$ 100
NRC	$ 16	$ 1
NSF	$ 1,154	$1,243
Smithsonian	$ 0	$ 25
Transportation	$ 99	$ 92
Total	**$ 2,891**	**$ 3,069**

Appendix Table A6:
STEM Education Funding in Millions by Agency[3]

---

[3] Only investments with a budget of $300,000 or greater in at least one year from FY 2012 to FY 2014 are included in the table. Basic information on investments with a budget consistently below $300,000 was and will continue to be collected as part of the annual NSTC Federal STEM Education Inventory. Investments that appear in the FY 2011 inventory (Appendix Table A4) and do not appear in the table were either eliminated in FY 2012 or fell below $300,000 a year.

Agency	Sub-Agency	Investment	FY 12 Enacted	FY 14 Requested
Agriculture	NIFA	1890 Facilities Grant Program	19.7	19.7
Agriculture	NIFA	1890 Institutions Capacity Building Grants Program: Extension	6.4	6.4
Agriculture	NIFA	1890 Institutions Capacity Building Grants Program: Teaching	6.4	6.4
Agriculture	NIFA	4-H Science, 4-H Youth Development Program	24.0	24.0
Agriculture	APHIS	AgDiscovery	0.5	0.5
Agriculture	NIFA	Agriculture in the Classroom	0.6	-
Agriculture	NIFA	AITC Secondary Postsecondary Agriculture Education Challenge Grants (SPECA)	0.9	-
Agriculture	NIFA	Alaska Native-Serving and Native Hawaiian-Serving Institutions Education Competitive Grants Program	3.2	3.2
Agriculture	NIFA	Distance Education Grants for Institutions of Higher Education in Insular Areas (DEG)	0.8	-

Agency	Sub-Agency	Investment	FY 12 Enacted	FY 14 Requested
Agriculture	NIFA	Food and Agricultural Sciences National Needs Graduate and Postgraduate Fellowship Grant Program	3.2	-
Agriculture	NIFA	Higher Education Challenge Grants (HEC)	4.8	-
Agriculture	NIFA	Higher Education Multicultural Scholars Program (MSP)	1.0	-
Agriculture	NIFA	Hispanic Serving Institutions Education Grants Program	9.2	9.2
Agriculture	NIFA	Insular Programs	-	1.7
Agriculture	NIFA	NIFA Fellowship Grants Program	6.1	13.7
Agriculture	NIFA	Resident Instruction Grants Program for Institutions of Higher Education in Insular Areas	0.9	-
Agriculture	NIFA	Women and Minorities in Science, Technology, Engineering and Mathematics Fields Program (WAMS)	0.4	-
Agriculture Total			88.1	84.8

Agency	Sub-Agency	Investment	FY 12 Enacted	FY 14 Requested
Commerce	NOAA	Competitive Education Grants (including Environmental Literacy Grants)	5.1	-
Commerce	NOAA	Coral Reef Conservation Program	0.5	-
Commerce	NOAA	Dr. Nancy Foster Scholarship Program	0.5	-
Commerce	NOAA	Educational Partnership Program with Minority Serving Institutions	12.5	14.4
Commerce	NOAA	Ernest F. Hollings Undergraduate Scholarship Program	4.9	5.4
Commerce	NOAA	National Estuarine Research Reserve System	0.6	-
Commerce	NOAA	National Sea Grant College Program - Education	0.8	-
Commerce	NIST	NIST Summer Institute for Middle School Teachers	0.3	-
Commerce	NOAA	NOAA Bay Watershed Education and Training (B-WET)	5.5	-

Agency	Sub-Agency	Investment	FY 12 Enacted	FY 14 Requested
Commerce	NOAA	NOAA Fisheries Education Program*	3.5	-
Commerce	NOAA	NOAA Office of Ocean Exploration and Research (Education Only)	0.9	-
Commerce	NOAA	NOAA Teacher at Sea Program	0.6	-
Commerce	NOAA	Satellite and Information Service*	3.2	-
Commerce	NIST	STEM Pipeline for the Next Generation Scientists and Engineers.	0.4	1.0
Commerce	NIST	Summer Undergraduate Research Fellowship (SURF)	0.8	0.8
		*FY 2012 figures are estimates.		
Commerce Total			**40.0**	**21.6**
Defense		Army Educational Outreach Program (AEOP)	8.2	9.4
Defense		Awards to Stimulate and Support Undergraduate Research Experiences (ASSURE)	4.5	4.5

Agency	Sub-Agency	Investment	FY 12 Enacted	FY 14 Requested
Defense		DoD STARBASE Program	25.0	-
Defense		Navy Historically Black Colleges and Universities/Minority Institutions Research and Education Partnership	1.4	1.5
Defense		Iridescent Learning	2.5	-
Defense		National Defense Education Program (NDEP) K-12 component	16.6	-
Defense		National Defense Education Program (NDEP) Science, Mathematics And Research for Transformation (SMART)	43.3	48.7
Defense		National Defense Science and Engineering Graduate (NDSEG) Fellowship Program	39.7	38.0
Defense		National Science Center (NSC)	1.9	-
Defense		Navy - Science and Engineering Apprenticeship Program (SEAP)	0.8	0.7
Defense		SeaPerch	1.5	1.5

Agency	Sub-Agency	Investment	FY 12 Enacted	FY 14 Requested
Defense	NSA	Stokes Educational Scholarship Program	2.0	1.8
Defense		The Naval Research Enterprise Intern Program (NREIP)	1.3	0.9
Defense		Uniformed Services University of the Health Sciences (USUHS)	0.5	-
Defense		University Laboratory Initiative (ULI)	2.2	-
Defense		University NanoSatellite Program	1.2	1.2
Defense Total			**152.6**	**108.2**
Education	OPE	Developing Hispanic Serving Institutions STEM and articulation programs	100.0	100.0
Education	OII	Fund for the Improvement of Education (FIE)	-	29.7
Education	OPE	Graduate Assistance in Areas of National Need (GAANN)	30.9	30.9

FEDERAL SCIENCE, TECHNOLOGY, ENGINEERING, AND MATHEMATICS (STEM) EDUCATION STRATEGIC PLAN

Agency	Sub-Agency	Investment	FY 12 Enacted	FY 14 Requested
Education	IES	High School Longitudinal Study of 2009	6.7	3.0
Education	OESE	Improving Teacher Quality State Grants/Effective Teacher and Leader State Grants	-	-
Education	OII	Investing in Innovation	28.5	45.0
Education	OESE	Mathematics and Science Partnerships/Effective Teaching and Learning for a Complete Education	149.7	149.7
Education	OPE	Minority Science and Engineering Improvement Program	9.5	9.5
Education	IES	Research in Special Education	3.3	6.5
Education	IES	Research, Development, and Dissemination	31.2	45.0
Education	OESE	STEM Innovation Networks	-	265.0
Education	OPE	Strengthening Predominantly Black Institutions	5.7	5.7

Agency	Sub-Agency	Investment	FY 12 Enacted	FY 14 Requested
Education	OESE	Teacher Incentive Fund	40.3	-
Education	OPE	Teacher Loan Forgiveness	79.2	80.0
Education	OPE	Upward Bound Math and Science Program	43.8	43.8
Education Total			**528.7**	**813.8**
Energy		Office of Energy Efficiency and Renewable Energy, Vehicle Technologies	Advanced Vehicle Competitions	2.0	2.0
Energy		Office of Science, Office of Nuclear Physics and Office of Basic Energy Sciences	American Chemical Society Summer School in Nuclear and Radiochemistry	0.6	-
Energy		Office of Science, Office of Workforce Development for Teachers & Scientists	Community College Internship (formerly Community College Institute of Science and Technology)	0.6	0.7
Energy		Office of Science, Advanced Scientific Computing Research	Computational Sciences Graduate Fellowship	6.0	-

Agency	Sub-Agency	Investment	FY 12 Enacted	FY 14 Requested
Energy	Office of Science, Office of Workforce Development for Teachers & Scientists	Visiting Faculty Program (formerly Faculty and Student Teams)	1.2	1.3
Energy	Office of Science, Office of Biological and Environmental Research	Global Change Education Program	0.6	-
Energy	Office of Energy Efficiency and Renewable Energy, Vehicle Technologies	Graduate Automotive Technology Education	1.0	-
Energy	Office of Environmental Management	HBCU Mathematics, Science & Technology, Engineering and Research Workforce Development Program	8.3	8.0
Energy	Office of Energy Efficiency and Renewable Energy, Industrial Technologies	Industrial Assessment Centers	3.7	6.0
Energy	Office of Economic Impact and Diversity	Minority Educational Institution Student Partnership Program	1.2	1.1
Energy	Office of Energy Efficiency and Renewable Energy, Solar Energy Technologies	Minority University Research Associates Program (MURA)	0.5	0.5

Agency	Sub-Agency	Investment	FY 12 Enacted	FY 14 Requested
Energy	Office of Science, Office of Workforce Development for Teachers and Scientists	Office of National Science Bowl	2.7	2.8
Energy	Office of Science, Office of Fusion Energy Sciences	Office of National Undergraduate Fellowship Program in Plasma Physics and Fusion Energy Sciences	0.4	-
Energy	Office of Science, Office of Workforce Development for Teachers and Scientists	Office of Graduate Student Research Program (formerly Office of Science Graduate Fellowship)	5.0	2.0
Energy	Office of Science, Office of Fusion Energy Sciences	Office of Plasma/Fusion Science Educator Programs	0.8	-
Energy	Office of Science, High Energy Physics	QuarkNet	0.6	-
Energy	Office of Science, Office of Workforce Development for Teachers and Scientists	Science Undergraduate Laboratory Internships	6.5	7.3
Energy	Office of Energy Efficiency and Renewable Energy, Building Technologies	Solar Decathlon	4.2	2.0
Energy	Office of Fossil Energy	Special Recruitment Programs/Mickey Leland Fellowship	0.7	0.7

Agency	Sub-Agency	Investment	FY 12 Enacted	FY 14 Requested
Energy	Office of Energy Efficiency and Renewable Energy, Wind Energy	Wind for Schools	0.9	-
Energy Total			47.5	34.4
EPA	ORD	Cooperative Training Partnership in Environmental Sciences Research	0.5	0.2
EPA	Office of Environmental Education	Environmental Education Grants	3.5	-
EPA	ORD	Greater Research Opportunities (GRO) Fellowships for Undergraduate Environmental Study	2.1	-
EPA	Office of Environmental Education	National Environmental Education and Training Partnership	2.0	-
EPA	ORD	P3-People, Prosperity & the Planet-Award: A National Student Design Competition for Sustainability	3.1	2.6
EPA	NCER	Science to Achieve Results Graduate Fellowship Program	14.0	-
EPA	ORD	University of Cincinnati/EPA Research Training Grant	0.6	0.6

Agency	Sub-Agency	Investment	FY 12 Enacted	FY 14 Requested
EPA Total			25.8	3.4
HHS	NIH, NIGMS	Bridges to the Baccalaureate Program	6.3	7.2
HHS	NIH, Intramural Training	Clinical Research Training Program	0.2	-
HHS	NIH, Office of Science Education	Curriculum Supplement Series	0.3	-
HHS	HRSA	Health Careers Opportunity Program	15.0	-
HHS	NIH, NIGMS	Initiative for Maximizing Student Development	2.7	2.2
HHS	NIH, NIGMS	MARC U-STAR NRSA Program	18.2	18.0
HHS	NIH, NICHD	Mathematics and Science Cognition and Learning (MSCL) Program	3.7	3.9
HHS		Medical Research Scholars Program (MRSP)	-	0.2

97

Agency	Sub-Agency	Investment	FY 12 Enacted	FY 14 Requested
HHS	NIH, NCI	National Cancer Institute Cancer Education and Career Development Program	0.9	1.1
HHS	NIH, NIAID	Science Education Awards	1.1	-
HHS	NIH, NINDS	Diversity Research Education Grants in Neuroscience	2.7	-
HHS	NIH, NLM	Institutional Grants for Research Training in Biomedical Informatics	0.2	-
HHS	NIH, OD	Science Education Partnership Award	15.4	-
HHS	NIH, Office of Science Education	Office of Science Education K-12 Program	2.2	-
HHS	HRSA	Public Health Traineeship	1.6	-
HHS	NIH, NIGMS	RISE (Research Initiative for Scientific Enhancement)	6.3	6.2
HHS	NIH	Ruth L. Kirschstein National Research Service Award Institutional Research Training Grants (T32, T35)	487.1	487.1

Agency	Sub-Agency	Investment	FY 12 Enacted	FY 14 Requested
HHS	NIH	Ruth L. Kirschstein NRSA for Individual Predoctoral Fellows, including Underrepresented Racial/Ethnic Groups, Students from Disadvantaged Backgrounds, and Predoctoral Students with Disabilities	4.0	3.7
HHS	NIH, NIDA	Science Education Drug Abuse Partnership Award	3.6	-
HHS	NIH, NICHD	Short Courses in Population Research (Education Programs for Population Research R25)	0.1	0.8
HHS	NIH, NIGMS	Short Courses on Mathematical, Statistical, and Computational Tools for Studying Biological Systems	0.5	-
HHS	NIH, NIEHS	Short Term Educational Experiences for Research (STEER) in the Environmental health Sciences for Undergraduates and High School Students	0.4	-
HHS	NIH, NHLBI	Short-Term Research Education Program to Increase Diversity in Health-Related Research	2.9	0.1
HHS	NIH, Intramural Training	Student Intramural Research Training Award Program	0.2	0.2
HHS	NIH, NHLBI	Summer Institute for Training in Biostatistics	0.1	0.1

Agency	Sub-Agency	Investment	FY 12 Enacted	FY 14 Requested
HHS	NIH, Intramural Training	Undergraduate Scholarship Program for Individuals from Disadvantaged Backgrounds	0.1	0.1
HHS Total			**575.7**	**530.8**
Homeland Security	S&T Office of University Programs	Homeland Security STEM Career Development Grant Program	2.7	-
Homeland Security	DNDO	National Nuclear Forensics Expertise Development Program	5.5	5.6
Homeland Security	S&T Office of University Programs	Scientific Leadership Awards Program	2.9	3.0
Homeland Security Total			**11.1**	**8.6**
Interior	Bureau of Land Management	Conservation and Land Management Internship Program	1.6	1.5
Interior	USGS	Educational Component of the National Geologic Mapping Program (EDMAP)	0.6	0.6
Interior	National Park Service	George Melendez Wright Climate Change Youth Initiative	0.4	0.4
Interior	National Park Service	Geoscientists-in-the-Parks Program	0.3	0.2

Agency	Sub-Agency	Investment	FY 12 Enacted	FY 14 Requested
Interior Total			3.0	2.7
NASA	ARMD	Aeronautics Academy	0.5	-
NASA	ARMD	Aeronautics Content - Smart Skies/Product Content Upgrade	0.8	-
NASA	ARMD	Aeronautics Scholarship	1.8	-
NASA	Education Office	AESP - Aerospace Education Services Project	3.1	-
NASA	Science Mission Directorate (SMD)	Aqua	0.3	-
NASA	Science Mission Directorate (SMD)	Astrophysics Forum	1.0	-
NASA	Science Mission Directorate (SMD)	Aura	0.3	-
NASA	Science Mission Directorate (SMD)	Cassini	0.9	-
NASA	Center JSC	CEP - Career Exploration Project	1.1	-

Agency	Sub-Agency	Investment	FY 12 Enacted	FY 14 Requested
NASA	Science Mission Directorate (SMD)	Chandra	0.5	-
NASA	Center JPL	Curriculum Improvement Partnership Award for the Integration of Research into the Undergraduate Curriculum (CIPAIR)	1.6	-
NASA	Science Mission Directorate (SMD)	DAWN	0.3	-
NASA	ARMD	Design Competitions	0.1	-
NASA	Science Mission Directorate (SMD)	Earth Science E/PO Forum	0.9	-
NASA	Center MSFC	eEducation Small Projects/Central Operation of Resources for Educators (CORE)	0.7	-
NASA	Center JSC	EFP - Education Flight Projects	2.2	-
NASA	Science Mission Directorate (SMD)	EPOESS	6.6	-
NASA	ESMD	ESMD Space Grant Project	0.6	-

Agency	Sub-Agency	Investment	FY 12 Enacted	FY 14 Requested
NASA	Education Office	GCCE - Global Climate Change Education	3.2	-
NASA	Science Mission Directorate (SMD)	GLOBE Program	4.5	4.5
NASA	Science Mission Directorate (SMD)	GRAIL	0.4	-
NASA	Education Office	GSRP - Graduate Student Researchers Program	2.8	-
NASA	Science Mission Directorate (SMD)	Heliophysics E/PO Forum	0.7	-
NASA	SOMD	HEOMD-Goldstone Apple Valley Radio Telescope (GAVRT) Project	0.5	-
NASA	ESMD	HEOMD-NASA's Beginning Engineering, Science and Technology (BEST) Students (NBS)	0.4	-
NASA	ESMD	HEOMD-University Student Launch Initiative	0.3	-
NASA	Science Mission Directorate (SMD)	HST	1.6	-

Agency	Sub-Agency	Investment	FY 12 Enacted	FY 14 Requested
NASA	Education Office	Informal STEM Education	10.0	-
NASA	ARMD	Innovation in Aeronautics Instruction Competition	0.1	-
NASA	Education Office	Innovation in Higher Education STEM Education	0.5	-
NASA	Education Office	INSPIRE - Interdisciplinary National Science Program Incorporating Research and Education Experience	0.7	-
NASA	Education Office	JPFP - Jenkins Pre-Doctoral Fellowship Program	2.6	-
NASA	Science Mission Directorate (SMD)	Juno	0.6	-
NASA	Center LaRC	LARSS - NASA Langley Aerospace Research Summer Scholars Program	0.6	-
NASA	Science Mission Directorate (SMD)	LDCM	0.6	-
NASA	Education Office	LEARN - Learning Environment and Research Network	3.0	-

Agency	Sub-Agency	Investment	FY 12 Enacted	FY 14 Requested
NASA	Center GRC	LERCIP - Lewis Educational Research Collaborative Internship Project (College)	0.1	-
NASA	Education Office	LTP - Learning Technologies Project	0.5	-
NASA	Science Mission Directorate (SMD)	Mars E/PO Formal Ed	1.1	-
NASA	Science Mission Directorate (SMD)	Mars E/PO Informal Ed	1.0	-
NASA	Science Mission Directorate (SMD)	MESSENGER	0.3	-
NASA	Education Office	MUREP (4 STEM programs in FY 2012 Inventory: MUREP Small Projects, NASA Science and Technology Institute for Minority Institutions, NASA Tribal College and University Project, University Research Centers)	18.8	30.0
NASA	Center GRC	MUST - Motivating Undergraduates in Science and Technology	2.3	-
NASA	Center JSC	NAS - NASA Aerospace Scholars	0.3	-
NASA	Education Office	NES - NASA Explorer Schools	3.8	-

Agency	Sub-Agency	Investment	FY 12 Enacted	FY 14 Requested
NASA	Center MSFC	NETS - NASA Education Technologies Services	1.0	-
NASA	ESMD	NSBRI Higher Education Activities - National Space Biomedical Research Institute	0.8	-
NASA	Science Mission Directorate (SMD)	Planetary Science E/PO Forum	0.9	-
NASA	SOMD	Reduced Gravity Student Flight Opportunity Project	0.3	-
NASA	Education Office	Research Cluster	1.4	-
NASA	ESMD	SEED - Systems Engineering Educational Discovery	0.3	-
NASA	Center GRC	SEMAA - Science Engineering Mathematics and Aerospace Academy/FIRST Buckeye	1.0	-
NASA	Science Mission Directorate (SMD)	SOFIA (Stratospheric Observatory for Infrared Astronomy) Education and Public Outreach	0.6	-
NASA	Education Office	SOI - Summer of Innovation/NASA IV&V Engineering Apprenticeship Program	5.9	-

Agency	Sub-Agency	Investment	FY 12 Enacted	FY 14 Requested
NASA	Education Office	Space Grant - National Space Grant College and Fellowship Program	40.0	24.0
NASA	OCT-ST	Space Technology Research Fellowships	12.0	15.0
NASA	CMO/ARC	Spaceward Bound	0.4	-
NASA	Education Office	STEM Accountability and Coordination (formerly STEM Education & Accountability Project)	-	26.7
NASA	Education Office	USRP - Undergraduate Student Research Project	0.3	-
NASA Total			149.4	100.2
NRC	Office of the Chief Human Capital Officer	Integrated University Program	15.0	-
NRC	Small Business and Civil Rights Office	Minority Serving Institutions Program (MSIP)	0.7	0.7
NRC	Office of the Chief Human Capital Officer	Grants to Universities (Curriculum Development) Program	-	-
NRC Total			15.7	0.7

Agency	Sub-Agency	Investment	FY 12 Enacted	FY 14 Requested
NSF	Directorate for Education and Human Resources (EHR)	Advanced Informal STEM Learning (AISL), formerly Informal Science Education (ISE)	61.4	47.8
NSF	Directorate for Education and Human Resources (EHR)	Advanced Technological Education (ATE)	64.0	64.0
NSF	Directorate for Education and Human Resources (EHR)	Alliances for Graduate Education and the Professoriate (AGEP)	7.8	7.8
NSF	Directorate for Education and Human Resources (EHR)	Catalyzing Advances in Undergraduate STEM Education (CAUSE)	-	123.1
NSF	Directorate for Geosciences (GEO)	Centers for Ocean Sciences Education Excellence	4.2	0.9
NSF	Directorate for Education and Human Resources (EHR)	Climate Change Education (CCE)	10.0	-
NSF	Directorate for Computer & Information Science & Engineering (CISE)	Computing Education for the 21st Century (CE21)	15.0	-

Agency	Sub-Agency	Investment	FY 12 Enacted	FY 14 Requested
NSF	Office of Cyberinfrastructure (OCI)	Cyberinfrastructure Training, Education, Advancement, and Mentoring for Our 21st Century Workforce (CI-TEAM)	4.0	-
NSF	Directorate for Education and Human Resources (EHR)	Discovery Research K-12 (DR-K12)	99.2	102.5
NSF	Office of International & Integrative Activities (OIIA)	East Asia & Pacific Summer Institutes for U.S. Graduate Students (EAPSI)	2.4	2.4
NSF	Directorate for Engineering (ENG)	Engineering Education (EE)	11.1	-
NSF	Directorate for Math and Physical Sciences (MPS)	Enhancing the Mathematical Sciences Workforce in the 21st Century (EMSW21)	11.8	11.5
NSF	Directorate for Education and Human Resources (EHR)	Excellence Awards in Science and Engineering (EASE)	5.2	4.8
NSF	Directorate for Education and Human Resources (EHR)	Federal Cyber Service: Scholarship for Service (SFS)	45.0	25.0

Agency	Sub-Agency	Investment	FY 12 Enacted	FY 14 Requested
NSF	Directorate for Geosciences (GEO)	Geoscience Education	1.5	-
NSF	Directorate for Geosciences (GEO)	Geoscience Teacher Training (GEO-Teach)	2.0	-
NSF	Directorate for Geosciences (GEO)	Global Learning and Observations to Benefit the Environment (GLOBE)	1.1	-
NSF	Directorate for Education and Human Resources (EHR) & Office of International and Integrative Activities (OIIA)	Graduate Research Fellowship Program (GRFP)	198.1	325.1
NSF	Directorate for Education and Human Resources (EHR)	Graduate Teaching Fellows in K-12 Education (GK-12)	27.0	-
NSF	Directorate for Education and Human Resources (EHR)	Historically Black Colleges and Universities Undergraduate Program (HBCU-UP)	31.9	31.9
NSF	Directorate for Education and Human Resources (EHR)	Innovative Technology Experiences for Students and Teachers (ITEST)	25.0	25.0

Agency	Sub-Agency	Investment	FY 12 Enacted	FY 14 Requested
NSF	Directorate for Education and Human Resources (EHR)	Integrative Graduate Education and Research Traineeship (IGERT)	59.8	-
NSF	Office of International & Integrative Activities (OIIA)	International Research Experiences for Students (IRES)	3.2	2.3
NSF	Directorate for Education and Human Resources (EHR)	Louis Stokes Alliances for Minority Participation (LSAMP)	45.6	45.6
NSF	Directorate for Education and Human Resources (EHR)	Math and Science Partnership (MSP)	57.1	-
NSF	Directorate for Engineering (ENG)	Nanotechnology Undergraduate Education in Engineering	1.5	-
NSF	NSF	NSF Research Traineeships (NRT)	-	55.1
NSF	Directorate for Education and Human Resources (EHR)	NSF Scholarships in Science, Technology, Engineering, and Mathematics (S-STEM)	75.0	75.0

Agency	Sub-Agency	Investment	FY 12 Enacted	FY 14 Requested
NSF	Directorate for Geosciences (GEO)	Opportunities for Enhancing Diversity in the Geosciences	3.6	-
NSF	Directorate for Engineering (ENG)	Research Experiences for Teachers (RET) in Engineering and Computer Science	5.5	5.5
NSF	Directorate for Education and Human Resources (EHR)	Research Experiences for Undergraduates (REU)	66.0	79.2
NSF	Directorate for Education and Human Resources (EHR)	Research in Disabilities Education (RDE)	6.5	-
NSF	Directorate for Education and Human Resources (EHR)	Research on Education and Learning (REAL), formerly Research and Evaluation on Education in Science and Engineering (REESE)	37.7	60.4
NSF	Directorate for Education and Human Resources (EHR)	Research on Gender in Science and Engineering (GSE)	10.5	-
NSF	Directorate for Education and Human Resources (EHR)	Robert Noyce Scholarship (Noyce) Program	54.9	60.9

112

Agency	Sub-Agency	Investment	FY 12 Enacted	FY 14 Requested
NSF	Directorate for Education and Human Resources (EHR)	Science, Technology, Engineering, and Mathematics Talent Expansion Program (STEP)	25.3	-
NSF		STEM-C Partnerships	-	73.6
NSF	Directorate for Education and Human Resources (EHR)	Transforming Undergrad Education in STEM (TUES)	39.5	-
NSF	Directorate for Biological Sciences (BIO)	Transforming Undergraduate Biology Education (TUBE)	13.0	-
NSF	Directorate for Education and Human Resources (EHR)	Tribal Colleges and Universities Program (TCUP)	13.3	13.3
NSF	Directorate for Education and Human Resources (EHR)	Widening Implementation and Demonstration of Evidence-based Reforms (WIDER)	8.0	-
NSF Total			1,153.7	1,242.8
Smithsonian		STEM Informal Education and Instruction	-	25.0

Agency	Sub-Agency	Investment	FY 12 Enacted	FY 14 Requested
Smithsonian Total			-	25.0
Transportation	Federal Aviation Administration (FAA)	Air Transportation Centers of Excellence	12.5	12.5
Transportation	Research and Innovative Technology Administration (RITA)	Dwight David Eisenhower Transportation Fellowship Program	1.9	1.5
Transportation	Research and Innovative Technology Administration (RITA)	Garrett A. Morgan Technology and Transportation Education Program	1.1	0.9
Transportation	Federal Highway Administration (FHWA) - Office of Civil Rights	National Summer Transportation Institute Program (STI)	2.6	3.0
Transportation	Research and Innovative Technology Administration (RITA)	Summer Transportation Institute Program for Diverse Groups (STIPDG)	1.3	1.5
Transportation	Research and Innovative Technology Administration (RITA)	University Transportation Centers Program	80.0	72.5
Transportation Total			99.4	91.9

Agency	Sub-Agency	Investment	FY 12 Enacted	FY 14 Requested
Federal Total			2,890.7	3,069.0

Appendix B

Investment Design Principles

Applying these principles in the design and implementation of investments is intended to improve the prospects of accomplishing the investment's primary objective and enhance the potential for stronger cross-agency coordination. These principles represent what might be considered useful frameworks for to allow more Federal investments to be successful, based on currently available evidence and best practice. Agencies will be at a variety of stages in their capacity and readiness to enact these principles, and as agencies work to do so, new principles will emerge and the current set will be refined.

There are three categories of design principles:
- *general investment design principles* that may apply to all investments
- *design principles by objective* for each of the main CoSTEM investment objective categories
- *design principles for investments serving underrepresented groups*

Federal STEM education investments should consider applying *general investment design principles* and the appropriate *design principles by objective.* Investments that aim to support groups traditionally underrepresented in STEM education should look to incorporate the *design principles for investments serving underrepresented groups.*

Over the period of implementation of the Strategic Plan, agency staff should collaborate with other agencies or non-Federal STEM education experts about how to incorporate design principles into their planning for existing and new investments, and how to build the expertise and capacity to do so over time. The activity of conforming to design principles will evolve as programs build that expertise and capacity. Thus, as implementation of the Strategic Plan begins investments will be assessed based on their progress in increasing capacity to align with the design principles.

General Investment Design Principles

CoSTEM recommends that agencies create or regularly update logic models or theory-of-action documents, management plans, and evaluation strategies for their investments, and address how their investments incorporate the applicable design principles.

As appropriate, investments should have a logic model or explicit theory of action that describes:
- Clear overarching goals, specific investment objectives and measurable outcomes
- How the investment helps to fulfill the agency's mission and connects to agency education or STEM assets[145], and STEM education goals
- Results of needs assessments, stakeholder input, or environmental scans that helped shape the goals, objectives, and outcomes

- Alignment with evidence-based practices, promising practices of experienced professionals, and relevant education research
- Strategic partnering within the agency, or with other Federal agencies, education organizations, or stakeholders, or why partnerships are not appropriate
- How the investment activities are likely to advance the goals and objectives

As appropriate, investments should have a management plan that describes:
- Needed expertise, staffing plan and how expertise is assured, plans for professional development of staff to implement design principles, and assignment of accountability for outcomes
- Strategies for dissemination or scale-up of promising practices and lessons learned in the course of implementation
- Plans for budget allocation and cost-sharing
- Strategic partnering within the agency, or with other Federal agencies, education organizations, or stakeholders, or why partnerships are not appropriate
- Resources needed for evaluation and staffing

As appropriate, investments should have an evaluation strategy that describes:
- A prioritized list of evaluation questions and goals
- Alignment of evaluation with the agency's STEM education evaluation strategy or agency strategic performance plan
- Evaluation methods, design and measures that are appropriate for the questions to be addressed, the outcomes being measured, the type of investment, its stage of implementation, and other relevant factors
- Data to be gathered and methods of data collection
- Use of standards for program evaluation[146] or evaluation research design
- Plans for sharing results of evaluation and identifying promising approaches
- Formal mechanisms by which evaluation findings inform improvements in investments

STEM Education for Underrepresented Groups Principles

In order to increase the number of individuals from underrepresented groups that graduate with STEM degrees, STEM investments should:

- Be designed with input from and implemented with the participation of underrepresented groups, local community stakeholders, and other stakeholders as appropriate
- Draw upon, relate to, and be respectful of the interests, knowledge, practices, and culturally relevant STEM experiences of underrepresented groups and demonstrate an understanding of targeted communities based on research
- Take advantage of place-based or experiential learning opportunities and agency STEM professionals, facilities, technology, and data where appropriate, particularly for mission agencies

- Build sustained relationships between participants and STEM partners (including continuous tracking and mentoring of participants)
- Build capacity and develop a strategy or mechanism to build a robust STEM workforce pipeline

Design Principles by Primary Investment Objective

The Federal STEM Education Portfolio Review (NSTC, 2011) requests that agencies identify a "single primary objective" for each investment (p. 13). In addition to applying the General Investment Design Principles, and those for underrepresented groups if applicable, investments should also consider applying the design principles listed here according to primary objective.

Learning Investments and Engagement Investments

Learning Investments "develop STEM skills, practices, or knowledge of students or the public." (NSTC, 2011, p. 13) Engagement Investments are designed to increase learners' involvement and interest in STEM, inform their view of the value of STEM in their lives, or enhance their ability to participate in STEM. These investments will address learners both inside and outside of formal education systems, at all levels. A single set of design principles can apply to both types of investments. Learning or Engagement Investments should work to:

- Have activities articulate learning goals describing what students or the public should know and be able to do as a result of engaging with the education activity, as well as goals for building interest, developing identity, improving motivation, and other affective outcomes, as appropriate
- Build from and acknowledge the significant expertise and professional knowledge of designers and developers of learning materials and activities
- Incorporate expectations for pilot, laboratory, or field testing of materials and activities, and iterative improvement and redesign
- Acknowledge the policy and learning contexts in which materials and activities would be used, with attention to such matters as national, state, or local standards; local or state education agency criteria; and available scientific expertise or facilities
- Encourage use of evidence-based practices and draw on effective approaches (e.g., active learning, career assessment, mentoring, coaching, cohort or team experiences, peer learning, authentic science experiences, citizen science, experiential, hands-on, inquiry, or place-based learning)
- Expect that learning and affective outcomes will be assessed, if possible using common approaches or instruments, across activities in the investment
- Include, as appropriate, comprehensive and coherent support or interface materials (e.g., teacher manuals, aligned assessments, affiliated websites, leadership supports, online platforms, etc.)

Pre- and In-Service Educator and Leader Performance

Investments in this area are designed to "train or retain STEM educators (K-12 pre-service or in-service, postsecondary, and informal) and education leaders to improve their content knowledge and pedagogical skills." (NSTC, 2011, p. 13)

Investments in pre-service educator or leader preparation should:

- Anticipate changing contexts for STEM educators or leaders (e.g., student demographics, learning technologies, accessible new scientific discoveries and innovations)
- Acknowledge the appropriate policy context for educator or leader preparation (e.g., college- and career-readiness standards, teacher preparation accreditation, teacher evaluation, and state licensure and credentialing processes)
- Reflect understanding of research developments about educator preparation, including the role of STEM disciplinary knowledge, STEM knowledge for teaching, effective classroom practices in STEM education, clinical experiences and practices, and assessment of STEM learning

In addition, investments in in-service educator and leader development should:

- Expect that activities are connected to systems of in-service educator or leader support (e.g. instructional materials, workshops, assessments, coaching, administrative expertise in STEM education)
- Assess training needs for activity staff who interact with learners and provide professional development
- For educators, align with specific local or state education agency professional development systems or priorities or provide professional credentials
- Serve multiple teachers or leaders in a school

Postsecondary STEM Degrees, Workforce Development, and STEM Careers

Postsecondary STEM degree investments are aimed at increasing the number of students who enroll in STEM majors, complete STEM credential or degree programs, or are prepared to enter STEM careers or advanced education. *STEM career investments* comprise general and agency mission-specific workforce education programs to prepare people to enter the STEM workforce with training or certification. STEM-discipline-specific knowledge and skills are the primary focus of the education investment.

Investments in either category should:

- Address specific STEM workforce capacity and expertise needs based on evidence from labor market data or workforce planning information
- Build involvement of representatives from potential employment and workforce sectors in design
- Establish strategic recruitment approaches that reach diverse audiences
- Have projects include appropriate mentoring, educational and internship experiences, academic support, and follow-up opportunities for networking and professional connections
- Incorporate authentic agency or STEM industry work or related experience

In addition, scholarship or fellowship investments should:

- Demonstrate that financial incentives such as scholarships or fellowships provide value added in attracting and retaining students to the workforce, if applicable
- Ensure sufficient support so that students successfully complete STEM programs or degrees

Institutional Capacity

Several Federal STEM education investments are designed to support the advancement and development of STEM personnel, programs, and infrastructure. This occurs in universities, informal education institutions, state education agencies, and local education agencies.

Investments should:
- Identify the institutional accountability, policy and incentive systems that may need transformation to support the goal, and recognize the importance of institutional context
- Justify choice of target institutions and provide rationale for why investment in these institutions is likely to leverage broader impact
- Build demonstration of commitment by institutional leadership
- Clarify "change strategies" for making improvements
- Have an analysis of resource needs at the institutional level
- Create clear sustainability strategies to extend beyond funding period

Education Research and Development

Investments in this area support the development of evidence-based STEM education models and practices. Such investments should:

- Incorporate explicit expectations about such key elements of research as theoretical and empirical bias, study design, data sources, and plans for interpretation and dissemination
- Encourage applicants to employ shared evidence standards, including efforts by NSF and IES

Include mechanisms for gathering feedback from small-scale implementers and practitioners that informs more basic R&D, to ensure that findings from basic research inform program development and implementation, and to help with relevance and usability.

References

[1] http://www.whitehouse.gov/sites/default/files/docs/stem_teachers_release_3-18-13_doc.pdf

[2] Pub. L. No. 111-358 (http://www.gpo.gov/fdsys/pkg/BILLS-111hr5116enr/pdf/BILLS-111hr5116enr.pdf)

[3] Pub. L. No. 111-358 (http://www.gpo.gov/fdsys/pkg/BILLS-111hr5116enr/pdf/BILLS-111hr5116enr.pdf).

[4] Members of the FC-STEM are listed on p. iii

[5] Members of CoSTEM are listed on page iv. The same agencies were represented on the FI-STEM.

[6] NSTC (2011). The Federal STEM Education Portfolio. http://www.whitehouse.gov/sites/default/files/microsites/ostp/costem_Federal_stem_education_portfolio_report.pdf.

[7] STEM fields are defined in the National Science Foundation's Science and Engineering Indicators, http://www.nsf.gov/statistics

[8] PCAST President's Council of Advisors on Science and Technology. (February 2012). Report to the President: Engage to excel: Producing one-million additional college graduates with degrees in science, technology, engineering, and mathematics. http://www.whitehouse.gov/sites/default/files/microsites/ostp/pcast-engage-to-excel-final_2-25-12.pdf

[9] Fleischman, H.L., Hopstock, P.J., Pelczar, M.P., and Shelley, B.E. 2010. *Highlights From PISA 2009. Performance of U.S. 15-Year-Old Students in Reading, Mathematics, and Science Literacy in an International Context* (NCES 2011-004). U.S. Department of Education, National Center for Education Statistics.

[10] Committee on Underrepresented Groups and the Expansion of the Science and Engineering Workforce Pipeline. 2010. *Expanding Underrepresented Minority Participation: America's Science and Technology Talent at the Crossroads.* Committee on Science, Engineering, and Public Policy; Policy and Global Affairs; National Academy of Sciences, National Academy of Engineering, and Institute of Medicine. Available from http://www.nap.edu/catalog/12984.html.

[11] Economics and Statistics Administration. (2011). Women in STEM: A Gender Gap to Innovation. United States Department of Commerce, Washington, D.C.

[12] This fifth priority, focused on the STEM workforce, was discussed in CoSTEM and was not specified as a goal in the progress report of February 2012. It has been recast as a goal in light of the President's FY 2014 STEM education reorganization. It has since been added in the development of this final report.

[13] Economics and Statistics Administration. (2011). STEM: Good Jobs Now and for the Future. United States Department of Commerce, Washington, D.C.

[14] U.S. Department of Commerce (January, 2012). The competitiveness and innovative capacity of the United States. http://www.commerce.gov/sites/default/files/documents/2012/january/competes_010511_0.pdf.

[15] http://www9.georgetown.edu/grad/gppi/hpi/cew/pdfs/stem-complete.pdf

[16] Business-Higher Education Forum. (2007). An American Imperative Transforming the Recruitment, Retention, and Renewal of Our Nation's Mathematics and Science Teaching Workforce. Retrieved from http://www.eric.ed.gov/PDFS/ED503709.pdf.

[17] Tai, R. H., Liu, Q. C., Maltese, A. V., & Fan, X. (2006). Planning early for careers in science. *Science.* 312, 1143 - 1144.

[18] Martin, M.O., Mullis, I.V.S., Foy, P., & Stanco, G.M. (2012). Chestnut Hill, MA: TIMSS & PIRLS International Study Center, Boston College.

[19] Carnevale, A.P., Smith, Nicole, and Melton, M. *STEM.* 2011. Georgetown University, Washington, D.C. http://www9.georgetown.edu/grad/gppi/hpi/cew/pdfs/STEMWEBINAR.pdf

[20] PCAST President's Council of Advisors on Science and Technology. (February 2012). Report to the President: Engage to excel: Producing one-million additional college graduates with degrees in science, technology,

engineering, and mathematics. http://www.whitehouse.gov/sites/default/files/microsites/ostp/pcast-engage-to-excel-final_2-25-12.pdf

[21] http://www.nsf.gov/statistics/seind12/pdf/c02.pdf

[22] National Research Council and National Academy of Engineering. *Community Colleges in the Evolving STEM Education Landscape: Summary of a Summit.* 2012. Washington, DC: The National Academies Press.

[23] National Center for Education Statistics. (2011). (Table Illustration Digest of Education Statistics May 3, 2013). *Number of persons age 18 and over, by highest level of educational attainment, sex, race/ethnicity, and age:2011.* Retrieved from http://nces.ed.gov/programs/digest/d11/tables/dt11_009.asp.

[24] http://nces.ed.gov/pubs2011/2011317.pdf

[25] Division of Science Resources Statistics. *Women, Minorities, and Persons with Disabilities in Science and Engineering: 2011.* Special Report NSF 11-309. Arlington, VA.

[26] http://www.whitehouse.gov/sites/default/files/microsites/ostp/pcast-engage-to-excel-final_2-25-12.pdf

[27] Federal Coordinating Council for Science, Engineering, and Technology, Committee on Education and Human Resources (1993). Pathways to excellence: A Federal strategy for science, mathematics, engineering, and technology education. http://www.eric.ed.gov/PDFS/ED360165.pdf.

[28] NSTC Subcommittee on Education (2008). Finding out what works: Agency efforts to strengthen the evaluation of Federal STEM education programs. http://www.whitehouse.gov/files/documents/ostp/NSTCpercent20Reports/NSTC_Education_Report_Complete.pdf .

[29] National Research Council. (2008). *NASA's Elementary and Secondary Education Program: Review and Critique.* Committee for the Review and Evaluation of NASA's Precollege Education Program, Helen R. Quinn, Heidi A. Schweingruber and Michael A. Feder, Editors. Board on Science Education. Washington, DC: The National Academies Press.

[30] National Research Council. (2010). *NOAA's Education Program: Review and Critique.* Committee for the Review of the NOAA Education Program. J.W. Farrington and M.A. Feder, Editors. Board on Science Education. Washington, DC: The National Academies Press.

[31] http://www2.ed.gov/programs/racetothetop/index_html

[32] http://www2.ed.gov/programs/innovation/index.html

[33] Clotfelter, C.T., Ladd, H.F., and Vigdor, J.L. (2007). Teacher credentials and student achievement in high school: A cross-subject analysis with student fixed effects. *Economics of Education Review, 26*(6), 673-782.

[34] Rivkin, S.G. (2007). *Value added analysis and education policy.* Washington, DC: Urban Institute, Center for the Analysis of Longitudinal Data in Education Research.

[35] Boyd, D., Grossman, P., Lankford, H., Loeb, S., Rockoff, J., and Wyckoff, J. (2008). The narrowing gap in New York City: Teacher qualifications and its implications for student achievement in high-poverty schools. *Journal of Policy Analysis and Management, 27*(4), 793-818.

[36] Rockoff, J.E. (2004). The impact of individual teachers on student achievement: Evidence from panel data. *American Economic Review, 94*(2), 247-252.

[37] Hanushek, E. A. & Rivkin, S.G. The Distribution of Teacher Quality and Implications for Policy. *Annual Review of Economics, 4,* September 2012, pp. 131-157.

[38] Hanushek, Eric A. The Economic Value of Higher Teacher Quality. *Economics of Education Review, 30*(3), June 2011, pp. 466-479.

[39] Ingersoll, R., & Perda, D. (2010a). How high is teacher turnover and is it a problem? Philadelphia: University of Pennsylvania, Consortium for Policy Research in Education.

[40] Ingersoll, R.M. and Perda, D. (2010b). Is the supply of Mathematics and Science Teachers Sufficient? *American Educational Research Journal, 47*(3), 563-594.

[41] Tai, R. H., Liu, Q. C., Maltese, A. V., & Fan, X. (2006). Planning early for careers in science. Science. 312, 1143 - 1144.

[42] Bybee, R. W., Taylor, J., Gardner, A., Scotter, P., Powell, J., Westbrook, A. et al. (2006). *The BSCS 5E Instructional Model: Origins, Effectiveness, and Application.* Colorado Springs, CO: Commissioned Paper, National Institutes of Health, Office of Science Education.

[43] *Many Experts, Many Audiences: Public Engagement with Science and Informal Science Education*, with Ellen McCallie, Larry Bell, Tiffany Lohwater, John H. Falk, Brice V. Lewenstein, Cynthia Needham, and Ben Wiehe, A CAISE Inquiry Group Report, 2009.

[44] National Research Council and National Academy of Engineering, *Community Colleges in the Evolving STEM Landscape: Summary of a Summit*, 2012.

[45] Carnevale, A.P., N.Smith, and J. Strohl. (2010). Help Wanted: Projections of Jobs and Education Requirements through 2018. Washington, DC: Georgetown University Center on Education and the Workforce.

[46] Lacey, T. A. and B. Wright. (2009). "Occupational employment projections to 2018." Monthly Labor Review 132(11):82-123.

[47] Langdon, D., G. McKittrick, D. Beede, B. Khan, and M. Doms. (2011). "STEM: Good Jobs Now and for the Future." ESA Issue Brief #03-11. Washington, DC: U.S. Department of Commerce.

[48] U. S. Census Bureau (2011a), Population Division, "Table 3. Annual Estimates of the Resident Population by Sex, Race, and Hispanic Origin for the United States: April 1, 2000 to July 1, 2009 (NC-EST2009-03)."

[49] U.S. Census Bureau (2011b), Population Estimates,http://www.census.gov/popest .

[50] U.S. Census Bureau (2009). U.S. Population Projections. Retrieved May 24, 2012 from http://www.census.gov/population/projections html .U.S. Census, 2009

[51] http://pathwaysreport.org/rsc/pdf/ex_summary.pdf

[52] http://acd.od.nih.gov/biomedical_research_wgreport.pdf, http://www.fgereport.org/rsc/pdf/CFGE_report.pdf, http://www.nap.edu/rdp/ http://portal.acs.org/portal/PublicWebSite/about/governance/CNBP_031603

[53] This model, proposed in the President's FY 2013 Budget and included in the Senate appropriations bill for Labor/HHS/Education for programs serving disconnected youth, allows multiple Federal programs to blend funds to support outcome-focused strategies and to waive statutory requirements that are not necessary to meet outcome goals or ensure appropriate use of funds.

[54] Clotfelter, C.T., Ladd, H.F., and Vigdor, J.L. (2007). Teacher credentials and student achievement in high school: A cross-subject analysis with student fixed effects. *Economics of Education Review, 26*(6), 673-782.

[55] Rivkin, S.G. (2007). *Value added analysis and education policy*. Washington, DC: Urban Institute, Center for the Analysis of Longitudinal Data in Education Research.

[56] Boyd, D., Grossman, P., Lankford, H., Loeb, S., Rockoff, J., and Wyckoff, J. (2008). The narrowing gap in New York City: Teacher qualifications and its implications for student achievement in high-poverty schools. *Journal of Policy Analysis and Management, 27*(4), 793-818.

[57] Rockoff, J.E. (2004). The impact of individual teachers on student achievement: Evidence from panel data. *American Economic Review, 94*(2), 247-252.

[58] National Center for Education Statistics. Accessed on January 3, 2013 at http://nces.ed.gov/programs/projections/projections2020/tables/table_16.asp

[59] (Council of Chief State School Officers, Washington, DC, 2007)

[60] National Center for Alternative Certification, Accessed at http://www.teach-now.org/intro.cfm on April 22, 2013

[61] Ingersoll, R. and Perda, D. (2010). Is the Supply of Mathematics and Science Teachers Sufficient?, *American Educational Research Journal*, Vol. 43(3). pp. 563-594.

[62] Hess, F., Kelly, A., and Meeks, O. (2011). The Case for Being Bold A New Agenda for Business in Improving STEM Education. http://www.aei.org/papers/education/the-case-for-being-bold/ (April 2011).

[63] Schmidt et al.(2010). Breaking the Cycle: An International Comparison of U.S. Mathematics Teacher Preparation, Michigan State University Center for Research in Mathematics and Science Education.

[64] [64] Ball, D., Hill, H., and Bass, H. (2005). Knowing Mathematics for Teaching. *American Educator*. Fall 2005. pp. 14-22, 43-46.

[65] National Research Council. *Preparing Teachers: Building Evidence for Sound Policy*. Washington, DC: The National Academies Press, 2010.

[66] http://www.whitehouse.gov/sites/default/files/microsites/ostp/pcast-stemed-report.pdf, 59.

[67] Ball, D., Hill, H., and Bass, H. (2005). Knowing Mathematics for Teaching. *American Educator*. Fall 2005. pp. 14-22, 43-46.

[68] National Science and Technology Council, (December 2011). *The Federal Science, Technology, Engineering, and Mathematics (STEM) Education Portfolio. Washington, DC.*

[69] National Research Council. (2011). *STEM Smart Brief STEM Smart: Lessons Learned From Successful Schools.* Committee on Highly Successful Science Programs for K-12 Science Education. Board on Science Education and Board on Testing and Assessment, Division of Behavioral and Social Sciences and Education. Washington, DC: The National Academies Press.

[70] Ingersoll, R. and Perda, D. (2010). Is the Supply of Mathematics and Science Teachers Sufficient?, *American Educational Research Journal*, Vol. 43(3). pp. 563-594.

[71] Authentic STEM experiences may be provided to K-12 teachers via internships, fellowships, and scholarships; "grade-appropriate" acknowledges that the nature of authentic STEM experiences is likely to differ for PK-5, 6-8 and 9-12 grade educators.

[72] Authentic STEM experiences may be provided to P-12 teachers via internships, fellowships, and scholarships; "grade-appropriate" acknowledges that the nature of authentic STEM experiences is likely to differ for PK-5, 6-8 and 9-12 grade educators.

[73] L. Deslauriers, E. Schelew, C. Wieman, *Science* 332, 862 (2011).

[74] B. A. Nagda, S. R. Gregerman, J. Jonides, W. von Hippel, J. S. Lerner, *Rev. Higher Educ.* 22, 55 (1998).

[75] See, for example, Sivan, Leung, Woon & Kember (2000), An implementation of active learning and its effect on the quality of student learning. *Innovations in Education and Training International 37* (4): 381-389.

[76] Might the NRC Learning Science in Informal Settings report

[77] Knox, K. L., Moynihan, J. A., & Markowitz, D. G. (2003). Evaluation of Short-Term Impact of a High School Summer Science Program on Students' Perceived Knowledge and Skills. *Journal of Science Education & Technology, 12*(4), 471-478; Markowitz, D. G. (2004). Evaluation of the Long-Term Impact of a University High School Summer Science Program on Students' Interest and Perceived Abilities in Science. *Journal of Science Education & Technology, 13*(3), 395-407.

[78] http://www.nasa.gov/offices/education/programs/national/dln/special/DigitalBadges.html

[79] http://www.nationalservice.gov/newsroom/press-releases/2013/president-obama-announces-stem-americorps-inspire-young-peoples

[80] http://www.mnh.si.edu/education/yes/about.html

[81] http://us2020.org/

[82] Committee on STEM Education, *The Federal Science, Technology, Engineering and Mathematics Education Portfolio,* Washington, D.C. National Science Technology Council.

[83] Efforts should reflect the recent OSTP memorandum on Increasing Access to the Results of Federally Funded Scientific Research
http://www.whitehouse.gov/sites/default/files/microsites/ostp/ostp_public_access_memo_2013.pdf

[84] Carnevale, A.P., N.Smith, and J. Strohl. (2010). Help Wanted: Projections of Jobs and Education Requirements through 2018. Washington, DC: Georgetown University Center on Education and the Workforce.

[85] Lacey, T. A. and B. Wright. (2009). "Occupational employment projections to 2018." Monthly Labor Review 132(11):82-123.

[86] Langdon, D., G. McKittrick, D. Beede, B. Khan, and M. Doms. (2011). "STEM: Good Jobs Now and for the Future." ESA Issue Brief #03-11. Washington, DC: U.S. Department of Commerce.

[87] http://my-goals.performance.gov/sites/default/files/images/STEM%20Education%20CAP%20Goal%20-%20FY2012%20Quarter%204%20Update_2.pdf

[88] National Research Council. (2012). *Education for Life and Work: Developing Transferable Knowledge and Skills in the 21st Century.* Committee on Defining Deeper Learning and 21st Century Skills

[89] National Research Council and National Academy of Engineering, *Community Colleges in the Evolving STEM Landscape: Summary of a Summit,* 2012.

[90] National Research Council, *Discipline-Based Education Research: Understanding and Improving Learning in Undergraduate Science and Engineering,* 2012.

[91] PCAST (February 2012). Report to the President: Engage to excel: Producing one-million additional college graduates with degrees in science, technology, engineering, and mathematics. http://www.whitehouse.gov/sites/default/files/microsites/ostp/pcast-engage-to-excel-final_2-25-12.pdf

[92] http://cep.mit.edu

[93] http://my-goals.performance.gov/sites/default/files/images/STEM%20Education%20CAP%20Goal%20-%20FY2012%20Quarter%204%20Update_2.pdf

[94] http://www.nsf.gov/news/news_summ.jsp?cntn_id=127902&org=NSF&from=news

[95] Langdon, D.. McKittrick, G. Beede, D. Khan, B., Julian, T., Lehrman, R., and Doms, M. (2011). Education Supports Racial and Ethnic Equality in STEM. ESA Issue Brief #05-11. Washington, DC: U.S. Department of Commerce.

[96] Langdon, D., McKittrick, G., Beede, D., Khan, B., and Doms, M. (2011). Women in STEM: A Gender Gap to Innovation. ESA Issue Brief #04-11. Washington, DC: U.S. Department of Commerce.

[97] C. Loes *et al.,* Effects of diversity experiences on critical thinking skills over four years of college; www.education.uiowa.edu/centers/docs/cdre-documents/Loes_Pascarella_and_Umbach_2012_3.pdf?sfvrsn=0

[98] S. E. Page, The Difference: How the Power of Diversity Creates Better Groups, Firms, Schools, and Societies (Princeton Univ. Press, Princeton, NJ, 2007).

[99] U.S. Department of Education, National Center for Education Statistics, Integrated Postsecondary Education Data System (IPEDS), Fall 2010, Completions component (prepared November 2011), *available at* http://nces.ed.gov/programs/digest/d11/tables/dt11_301.asp.

[100] *ibid.*

[101] *ibid.*

[102] Wei, et al., "Science, Technology, Engineering, and Mathematics (STEM) Participation Among College Students with an Autism Spectrum Disorder," Journal of Autism and Developmental Disorders, November 1, 2012, Table 4, *available at* http://link.springer.com/article/10.1007/s10803-012-1700-z/fulltext.html

[103] The data from the CRDC, while covering 85% of the nation's public school students, are not intended to be an estimation of national data. All data in the CRDC are self-reported.

[104] Alvarado, C. & Dodds, Z Women in CS: an evaluation of three promising practices, Proceedings of the 41st ACM technical symposium on Computer science education, 2010

[105] Walton, G. M. & Spencer, S. J. (2009). Latent ability: Grades and test scores systematically underestimate the intellectual ability of negatively stereotyped students. *Psychological Science, 20,* 1132-1139.

[106] Walton, G. M., & Cohen, G. L. (2011). A brief social-belonging intervention improves academic and health outcomes of minority students. *Science,331*(6023), 1447-1451.

[107] *Employment Projections: 2010-2020,* Bureau of Labor Statistics, 2012.

[108] PCAST (February 2012). Report to the President: Engage to excel: Producing one-million additional college graduates with degrees in science, technology, engineering, and mathematics. http://www.whitehouse.gov/sites/default/files/microsites/ostp/pcast-engage-to-excel-final_2-25-12.pdf

[109] *Survey of Doctorate Recipients*, NSF/National Center for Science and Engineering Statistics, 2008.

[110] *Science and Engineering Indicators 2012* Arlington, VA (NSB 12-01) | January 2012

[111] http://grants nih.gov/grants/guide/pa-files/PA-11-184.html

[112] Investing in the Future: NIGMS Strategic Plan for Biomedical and Behavioral Research Training; Blueprint for Implementation. http://www nigms nih.gov/Training/StrategicPlanImplementationBlueprint.htm

[113] National Research Council. "Goals for U.S. STEM Education." *Successful K-12 STEM Education: Identifying Effective Approaches in Science, Technology, Engineering, and Mathematics*. Washington, DC: The National Academies Press, 2011.

[114] National Research Council. (2012). *Discipline-Based Education Research: Understanding and Improving In Undergraduate Science and Engineering*. Committee on the Status, Contributions, and Future Directions of Discipline-Based Education Research. Board on Science Education Washington, DC: The National Academies Press.

[115] http://ies.ed.gov/ncee/projects/evaluation/tq.asp

[116] Charles T. Clotfelter, Helen F. Ladd, and Jacob L. Vigdor (2013), *Algebra for 8th Graders: Evidence on its Effects from 10 North Carolina Districts*, working paper

[117] http://grants nih.gov/training/career_progress/index.htm

[118] National Research Council. (2007). *Taking Science to School: Learning and Teaching Science in Grades K-8.* Washington, DC: The National Academies Press.

[119] National Research Council. (2009). *Learning Science in Informal Environments: People, Places, and Pursuits.* Committee on Learning Science in Informal Environments. Philip Bell, Bruce Lewenstein, Andrew W. Shouse, and Michael A. Feder, Editors. Board on Science Education, Washington, DC: The National Academies Press.

[120] National Research Council. (2011). *Successful K-12 STEM Education: Identifying Effective Approaches in Science, technology, Engineering, and Mathematics*. Committee on Highly Successful Science Programs for K-12 Science Education. Board on Science Education and Board on Testing and Assessment. Washington, DC: The National Academies Press.

[121] National Research Council. (2011). *Successful K-12 STEM Education: Identifying Effective Approaches in Science, technology, Engineering, and Mathematics*. Committee on Highly Successful Science Programs for K-12 Science Education. Board on Science Education and Board on Testing and Assessment. Washington, DC: The National Academies Press.

[122] National Research Council. (2012). *Discipline-Based Education Research: Understanding and Improving In Undergraduate Science and Engineering*. Committee on the Status, Contributions, and Future Directions of Discipline-Based Education Research. Board on Science Education Washington, DC: The National Academies Press.

[123] National Research Council (2001). *Expanding Underrepresented Minority Participation: America's Science and Technology Talent at the Crossroads*. Committee on Science, Engineering, and Public Policy. Washington, DC: National Academy Press.

[124] National Mathematics Advisory Panel. (2008). Foundations for Success The Final Report of the National Mathematics Advisory Panel. Retrieved from http://www2.ed.gov/about/bdscomm/list/mathpanel/report/final-report.pdf.

[125] National Research Council. (2010). *Preparing teachers: Building evidence for sound policy.* Committee on the Study of Teacher Preparation Programs in the United States. Division of Behavioral and Social Sciences and Education. Washington, DC: The National Academies Press.

[126] President's Council of Advisors on Science and Technology. (September, 2010). Prepare and inspire: K-12 education in science, technology, engineering, and mathematics (STEM) for America's future. http://www.whitehouse.gov/sites/default/files/microsites/ostp/pcast-stemed-report.pdf.

[127] President's Council of Advisors on Science and Technology (February 2012). Report to the President: Engage to excel: Producing one-million additional college graduates with degrees in science, technology, engineering, and mathematics. http://www.whitehouse.gov/sites/default/files/microsites/ostp/pcast-engage-to-excel-final_2-25-12.pdf

[128] National Research Council. (2008). *NASA's Elementary and Secondary Education Program: Review and Critique.* Committee for the Review and Evaluation of NASA's Precollege Education Program, Helen R. Quinn, Heidi A.

[129] National Research Council. (2010). *NOAA's Education Program: Review and Critique.* Committee for the Review of the NOAA Education Program. J.W. Farrington and M.A. Feder, Editors. Board on Science Education. Washington, DC: The National Academies Press.

[130] Friedman, A. (Ed.). (March 12, 2008). Framework for Evaluating Impacts of Informal Science Education Projects [On-line]. Retrieved from http://insci.org/resources/Eval_Framework.pdf.

[131] http://findyouthinfo.gov/docs/Common%20Standards-Draft%2002-28-13_508_3-11-13_Author.pdf

[132] http://www.regulations.gov/#!documentDetail;D=NSF-2012-OTR-0002-0001

[133] http://www.whitehouse.gov/sites/default/files/omb/memoranda/2012/m-12-14.pdf

[134] Analytic support was provided by the Science & Technology Policy Institute, which is a federally funded research and development center (FFRDC) chartered by Congress (https://www.ida.org/stpi.php)

[135] http://www.whitehouse.gov/sites/default/files/microsites/ostp/costem__federal_stem_education_portfolio_report.pdf

[136] http://cahsi.cs.utep.edu/Portals/0/Resources/Literature/Report_of_the_AcademicCompetitiveness_Council.pdf

[137] http://www.afa.org/ProfessionalDevelopment/AerospaceEducation/ReportsandStudies/RS_PDFs/Inventory-Survey.pdf

[138] FY 2008-2011 data were collected

[139] The total funding amount in figures is noted only if it differs from the FY 2011 sum of $2,891 million.

[140] While only one primary objective could be selected, multiple secondary objectives could be selected.

[141] The categories in this table are not mutually exclusive and thus cannot be summed.

[142] "Adults" does not include investments for educators

[143] Partnership and Interagency Collaboration information was only collected for broader STEM Education investments.

[144] More than one evaluation design could be identified for each investment.

[145] STEM and education expertise, resources, networks, facilities, policies

[146] See, for example, American Evaluation Association program evaluation standards, http://www.eval.org